# MASS SPECTROMETRY

# ELLIS HORWOOD SERIES IN ANALYTICAL CHEMISTRY

*Series Editors:* Dr MARY MASSON, University of Aberdeen,
and Dr JULIAN F. TYSON, Amherst, USA
*Consultant Editors:* Prof. J. N. MILLER, Loughborough University of Technology, and
Dr R. A. CHALMERS, University of Aberdeen

# MASS SPECTROMETRY

E. CONSTANTIN
Professor at the Institute of Chemistry, University of Strasbourg

A. SCHNELL
Docteur Ingenieur, University of Strasbourg

*Translator*
Dr M. H. CHALMERS
Reading

*Translation edited and revised by*
Dr R. A. CHALMERS
University of Aberdeen
Dr A. PAPE
University of Strasbourg

**ELLIS HORWOOD**
NEW YORK LONDON TORONTO SYDNEY TOKYO SINGAPORE

This English edition first published in 1990
and Reprinted in 1991 by
**ELLIS HORWOOD LIMITED**
Market Cross House, Cooper Street,
Chichester, West Sussex, PO19 1EB, England

A division of
Simon & Schuster International Group
A Paramount Communications Company

This English edition is translated from the original French edition *Specgtrométrie de masse*,
published in 1986 by Tec & Doc, France, © the copyright holders
© English Edition, Ellis Horwood 1990

Typeset by Ellis Horwood Limited
Printed and bound in Great Britain
by Bookcraft (Bath) Limited, Midsomer Norton, Avon

---

British Library Cataloguing in Publication Data

---

Constantin, E.
Mass spectrometry.
1. Mass spectrometry
I. Title II. Schnell, A. III. Spectromètrie de masse. *English*
543.0873
ISBN 0–13–555525–6 (Library Edn.)
ISBN 0–13–553363–5 (Student Edn.)

---

Library of Congress Cataloging-in-Publication Data

---

Constantin, E.
[Spectromé de masse. English]
Mass spectrometry / E. Constantin, A. Schnell; translator, M. H. Chalmers.
p. cm. — (Ellis Horwood series in analytical chemistry)
Translation of: Spectrométrie de masse.
Supt. of Docs. no.: 543/.0873
ISBN 0–13–555525–6 (Library Edn.)
ISBN 0–13–553363–5 (Student Edn.)
1. Mass spectrometry. I. Schnell, A. II. Title. II. Series.
QD96.M3C6413   1990

90–39403
CIP

# Table of contents

# Preface to the French first edition

In publishing this work we have tried to fill, at least partially, a gap in the series of books in French on mass spectrometry.

We have asembled in condensed form the basic essentials and the fields of application. With the increasing interpenetration of different disciplines, the specialist in one must more and more often acquire a certain mastery of several others. We hope this book will give the reader a taste for the subject as well as a basis and point of departure for its further study.

Mass spectrometry was born when J. J. Thomson showed the existence of stable isotopes, in his studies on the displacement of charged particles in electric and magnetic fields. The name 'mass spectrometry' is due to Aston (1920), and the term 'mass spectrometer' was first used, in 1926, by Smythe and Mattauch. This discipline has been enriched by fundamental studies and applications in physical and analytical chemistry.

However, the full development we know today was reached only through the evolution of other techniques, such as the use of lasers and computers.

In its elementary form, mass spectrometry primarily serves for determination of atomic and molecular weights. It has now become possible to correlate the spectrum and structure of a compound and identify its chemical bonds. The theory of mass spectra has permitted explanation of the mechanisms of fragmentation, and identification of the factors governing the probability of formation of different fragments. The major objectives of analytical mass spectrometry include prediction and general knowledge of the form of a molecule after its ionization, knowledge of whether it will remain intact or be fragmented, and quantitative analyses.

Although analysis is its preponderant role, mass spectrometry has other applications, such as the study of molecular energy levels, ion–molecule reaction mechanisms, atmospheric processes etc.

We have tried, without going beyond our chosen limits, to mention a wide range of applications, but have restricted the treatment to the essentials, without a detailed description.

The book is intended, above all, for scientists and engineers working in research laboratories and industry. It could help those who wish to be brought up to date with new developments in mass spectrometry, since nowadays it is essential for scientists and engineers in university and industrial research laboratories to keep pace with progress in knowledge.

We also think it could be useful to students in higher education (undergraduates, polytechnic students etc.) and also to high-school pupils. It may also help teachers by providing them with information they have previously had difficulty in finding.

The bibliography at the end of the work lists books and articles of general interest which the reader may consult to obtain more complete information about subjects of particular interest.

In the present state of development of chemistry, it seems to be inconceivable that its teaching should not include the methods of mass spectrometry. A serious pedagogic effort is urgently needed at all levels to accord this discipline a place more fitting to its importance and role. The example of the Anglo-Saxon countries in this respect is very instructive.

The Authors

# 1

# Mass spectrometry

## 1.1 INTRODUCTION

The mass spectrometer is an instrument that serves for establishment of the molecular weight and structure of organic compounds, and the identification and determination of the components of inorganic substances.

The sample is volatilized within the spectrometer and gas-phase ions formed from it are separated according to their mass/charge ($m/z$) ratios, and are usually detected electrically. The ion-currents corresponding to the different species are amplified and either displayed on an oscilloscope or a chart-recorder, or are stored in a computer. An example of a mass spectrum thus obtained is shown in Fig. 1.1. The

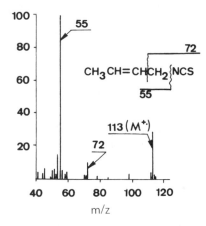

Fig. 1.1 — A typical mass spectrum, showing fragmentation pattern.

peak intensities are plotted as ordinates, in arbitrary units or normalized with respect to the most important peak (or some other selected peak), which is assigned a value of 100.

## 1.2 CONSTITUENT PARTS

A mass spectrometer consists of the following basic units:

(1) an ion source where ions are formed from the sample;
(2) an analyser which separates the ions according to their $m/z$ values;
(3) a detector which gives the intensity of the ion current for each species; the detector output can be displayed or stored, to yield the mass spectrum;
(4) electronics of power supply and control of the three units above;
(5) various pumping systems.

## 1.3 IONIZATION AND FRAGMENTATION

Various methods of ionization can be used, the choice depending on the physical state of the sample and the volatility and thermal stability of the material.

Electron-impact ionization gives satisfactory results for gas-phase molecules. For inorganic solids such as salts, thermal ionization, field desorption and laser desorption are used. Atom or ion bombardment is suitable for ionization of organic compounds of high molecular weight. Table 1.1 gives a résumé of some ionization methods commonly used.

**Table 1.1** — Ionization methods

| Gases | Liquids | Solids |
|-------|---------|--------|
| electron impact | volatilization, | thermal ionization |
| photon impact | and ionization | laser desorption/ionization |
| | as for gases | ion bombardment |
| | | atom bombardment |
| | | electric discharge |
| | | field desorption/ionization |

Electron impact ionization of an organic molecule M results in formation of a molecular ion $M^{+\cdot}$ by removal of an electron, to leave the charged radical:

$$M + e^- \rightarrow M^{+\cdot} + 2e^-$$

During the ionization, a certain amount of internal energy is transmitted to the molecule. This amount of energy is variable, and its magnitude determines whether one or more fragmentation reactions can occur. The reactions most likely to take place are those which have low activation energy and lead to relatively stable ionic or neutral species:

$$M^{+\cdot} \;\rightarrow\; \text{ionic fragment} \;+\; \text{neutral fragment}$$

$$\begin{array}{ccc}
\text{ionic fragment} & + & \text{neutral fragment} \\
(BC^{+\cdot}) & & (\text{radical, atom,} \\
\downarrow & & \qquad\text{molecule}) \\
\text{ionic fragment} & + & \text{neutral fragment} \\
\downarrow & & \\
\text{etc.} & &
\end{array}$$

The successive fragmentation reactions lead, in general, to ions and radicals consisting of a single atom or a group of atoms. It should be noted that in fragmentation the charge is located on the fragment with the lowest ionization potential. This in the fragmentation of $BC^{+\cdot}$ into $B^+$ and $C\cdot$ or into $B\cdot$ and $C^+$, if the ionization potential of B is lower than that of C, then the major fragmentation path will be into $B^+$ and $C\cdot$; similarly, for fragmentation of an ion $BC^+$, under the same conditions the most probable fragmentation would be into $B^+$ and C, not into B and $C^+$.

Stable fragments can also be formed by a change in the relative positions of the atoms in a molecule, followed by decomposition of the ion. These changes are called rearrangement reactions.

# 2

# Methods of ionization

In a mass spectrometer the ions are formed in the ionization chamber of a unit called the ion source. Several methods can be used to convert the initially neutral sample into an ionized species in the gas phase.

## 2.1 ELECTRON IMPACT IONIZATION

This is the oldest and most widely used method. The substance is volatilized into the ionization chamber, where its molecules are bombarded with electrons and transformed into positively charged ions:

$$A + e^- \rightarrow A^+(\text{excited}) + 2e^-$$
$$\downarrow$$
$$A^+(\text{final state})$$

The ion current, $I$, for the ions produced is given by

$$I = ai_e \rho \sigma l$$

where $i_e$ is the electron current, $\rho$ the sample pressure in the ionization chamber, $\sigma$ the ionization cross-section of the molecules, $l$ the effective path-length of the electrons in the chamber, and $a$ the efficiency of extraction of the ions. Figure 2.1 shows a typical mass spectrum obtained by electron-impact ionization.

## 2.2 PHOTON IONIZATION

In this method the ionizing agent is the photon. The photons are generated by high-power lamps or by lasers. The spectral range of the photons used varies from the ultraviolet to the infrared. The ionization processes depend on the photon energy,

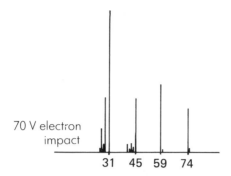

70 V electron
impact

31   45  59   74

Fig. 2.1 — Mass spectrum of diethyl ether obtained by electron impact. (Reproduced by permission, from C. A. McDowell (ed.), *Mass Spectrometry*, p. 93. Copyright 1963, McGraw-Hill Inc., New York.)

with either direct transition to the ionic state $A^+$ or multi-step photon interactions. Figure 2.2 shows a mass spectrum obtained by photon ionization. It is much simpler than the spectrum obtained for the same compound by electron impact ionization (Fig. 2.1).

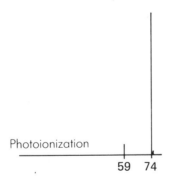

Photoionization

59   74

Fig. 2.2 — Mass spectrum of diethyl ether obtained by photon ionization. (Reproduced by permission, from C. A. McDowell (ed.), *Mass Spectrometry*, p. 93. Copyright 1963, McGraw-Hill Inc., New York.)

## 2.3  ION BOMBARDMENT

This method consists of producing secondary ions by subjecting the solid sample to bombardment by a primary ion beam. It is used for samples that have low volatility. The primary ions can have either very high energy (see Section 13.5, the californium

source) or low energy (SIMS sources: SIMS = secondary ion mass spectrometry).
Figure 2.3 shows the principle of SIMS, and Fig. 2.4 a typical mass spectrum.

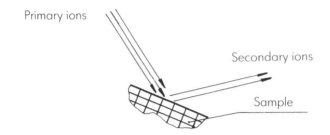

Fig. 2.3 — Principle of ionization by ion-bombardment.

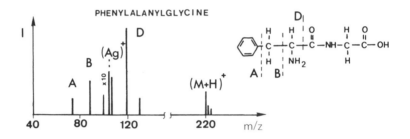

Fig. 2.4 — Mass spectrum obtained by SIMS. (Reproduced by permission, from A. Benninghoven and W. Sichtermann, *Org. Mass Spectrom.*, 1979, **12**, 595. Copyright 1979, Heyden Ltd.,
London.)

### 2.4  SPARK-SOURCE IONIZATION

This method is used for non-volatile inorganic solids (metals, minerals, salts, etc.).
Ion formation takes place in the electrode gap (D in Fig. 2.5) on spark discharge
between a conductor and an electrode made from a conductive sample, or between
two conductors, one of which carries a non-conductive sample.

### 2.5  BOMBARDMENT WITH NEUTRAL PARTICLES (Ar, Xe, Cs, etc.)

This method is also used for ionization of solid samples, and forms the basis of fast
atom bombardment (FAB, see Section 13.3). Figure 2.6 shows a typical mass
spectrum obtained by FAB.

### 2.6  IONIZATION BY ELECTROSPRAY

This technique is particularly useful for mass spectrometry of macromolecular
compounds such as polymers. A solution of the sample is forced through a fine metal

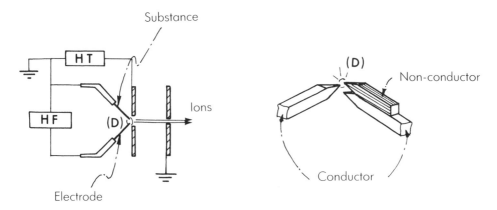

Fig. 2.5 — Principle of spark source ionization: (a) conducting sample; (b) non-conducting sample. HT=high tension; HF=high frequency.

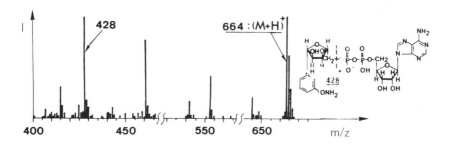

Fig. 2.6 — Mass spectrum obtained by fast-atom bombardment (FAB).

capillary tube (such as a hypodermic needle) that is at a high potential relative to the walls of the unit. The solution is converted into extremely small charged droplets, which vaporize in the ionization chamber.

## 2.7 THERMAL IONIZATION

The solid sample is placed on a filament, which when heated causes vaporization and ionization of the sample and thus the formation and emission of ions. The ratio of the number of ions formed $(n_+)$ to the number of neutral atoms vaporized $(n_0)$ depends on the temperature, the nature of the support, and the sample matrix.

### 2.7.1  Formation of positive ions

The efficiency of ionization, $n_+/n_0$, is given by the Langmuir–Saha equation:

$$n_+/n_0 = A \exp\left[(W-I)/kT\right]$$

where $W$ is the work function (energy needed to remove an electron from the surface) of the support material, $I$ is the first ionization energy of the sample element, $k$ is the Boltzmann constant, and $T$ is the absolute temperature of the support. $W$ and $I$ are expressed in eV. $A$ is constant at a given temperature, incorporating the internal reflection coefficients and statistical weights for the ions and neutral particles. The equation is valid if the sample is in thermal equilibrium with the support.

Since 1 eV is equivalent to $1.6021 \times 10^{-19}$ J, and the Boltzmann constant has the value $1.38054 \times 10^{-23}$ J/K, the equation may be written as

$$n_+/n_0 = A \exp\left[1.160 \times 10^4 (W-I)/T\right]$$

Clearly $n_+/n_0$ will be high if $W$ is greater than $I$ and small if it is not. The former is the case for caesium, potassium and rubidium on tantalum or tungsten supports, and the latter the case for most other combinations of element and support. For example, for silver ($I=7.57$ eV) on a tungsten support ($W=4.50$ eV), at 1030 K, assuming $A$ to be 0.5, $n_+/n_0$ would be $5 \times 10^{-16}$. For aluminium ($I=5.98$ eV) on a tungsten support at the same temperature, $n_+/n_0$ would be $3 \times 10^{-8}$.

Under certain circumstances, the sample may volatilize before the temperature needed to give adequate ionization has been reached, and a double or triple filament source may then be used, with volatilization from a low-temperature filament and ionization at a high-temperature filament. Advantages of the method are the small amount of sample needed, and the small energy spread of the ions, but a disadvantage is that isotopic fractionation occurs during volatilization of the sample.

### 2.7.2   Formation of negative ions

Negative ions may similarly be formed, but in this case the equation for the ionization efficiency is

$$n_-/n_0 = A \exp\left[(E_a - W)/kT\right]$$

where $E_a$ is the electron affinity of the element to be ionized, and $n_-$ is the number of negative ions produced.

## 2.8   FIELD IONIZATION

When a molecule in the gas phase passes through a strong electrostatic field (up to $10^8$ V/cm) it may lose an electron and yield a positively charged ion. The field is established by applying a potential difference of about 10 kV between a sharp metal edge (which may be activated by growth of an array of closely spaced microneedles) as the anode, and the cathode, which has an exit slit and is situated 0.5–2 mm away from the anode. *Field desorption* is a special case of this method, in which a solid sample is placed on the anode (an activated filament) of the electric field, and positive ions are ejected from the adsorbed layer. The mechanism of field desorption is still under debate.

Figure 2.7 shows a typical mass spectrum obtained by field ionization, and demonstrates the relative lack of fragmentation.

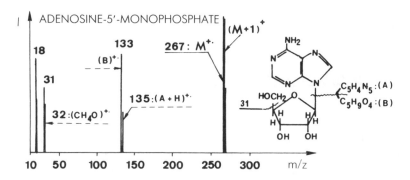

Fig. 2.7 — Mass spectrum obtained by field ionization. (Adapted from H. R. Schulten and H. D. Beckey, *Org. Mass Spectrom.*, 1973, **7**, 861. Copyright 1973, Heyden Ltd, London.)

## 2.9  CHEMICAL IONIZATION

In this method a reagent gas is introduced into the ionization chamber at about 1 torr pressure. When the sample is admitted (at a partial pressure of about 0.01 torr or less) and an electron beam is passed through the mixture, reagent gas ions are preferentially formed, and give ion–molecule reactions with the sample. This is a 'soft' method of ionization, resulting in little fragmentation. When the reagent ion is a stronger proton donor than $MH^+$, it produces an $MH^+$ ion from the sample molecule M. This system is discussed in detail in Chapters 11 and 13.

## 2.10  THERMOSPRAY IONIZATION

Although it is comparatively easy to couple a mass spectrometer to a gas chromatograph to act as the chromatographic detector, coupling to a liquid chromatograph is more difficult, because of the relatively large amount of solvent in the eluate. An early approach was to choose the eluent solvent so that when vaporized it could act as the reagent gas for chemical ionization, but this is rather limited in application. An alternative is to place the eluate on a moving belt or wire, which can then be heated to evaporate the solvent before electron impact, chemical or FAB ionization is applied directly to the sample on the wire or belt. However, when a solution containing an electrolyte is heated to about 150°C in a capillary and then sprayed into a vacuum, the solvent evaporates rapidly from the droplets and leaves solid particles consisting of solute molecules in a sheath of electrolyte. Because these particles are extemely small, the solute molecules are exposed to a very high ionic electrostatic field and are readily converted into protonated or cationized species (see Section 11.3.5), depending on the nature of the electrolyte. Either positive or negative ions can be produced in this way. This is a form of the technique known as thermospray (see page 111).

# 3

# Classical electron impact ionization source

This source (Figs. 3.1 and 3.2) consists of an evacuated ionization chamber (volume 1 cm$^3$) traversed by a beam of electrons, into which the sample is introduced in gaseous form. Inside the chamber there is a small charged plate called the repeller (R). Electrons produced from the heated filament (F) are accelerated by an electric field, and collimated into a narrow beam by a weak external magnetic field (not shown in Fig. 3.1). Collisions between the electrons and the sample molecules result in formation of positive ions (the number and type of which depend on the electron energy), and these are directed towards the exit slit (S in Fig. 3.2) by the potentials applied to the repeller plate and the extractor plate (1, in Fig. 3.1, which also focuses the ion beam). We can therefore speak of a quasi-optical system, in which the 'ion object' is situated in the electron beam, and the 'ion image' is formed at the entrance slit of the analyser. In effect, the inhomogeneous electric field between the electron beam and the analyser entrance corresponds to a medium of variable optical density (i.e. with a variable refractive index), capable of accelerating and focusing the ion beam. The focusing action of the field is due to the fact that the ion trajectories tend to be perpendicular to the equipotential lines, as shown in the 'ion optics' in Fig. 3.2.

It is therefore possible, by applying the proper potentials to the various plates, to obtain an ion beam that is narrow, intense, and sharply focused on the entrance to the analyser. Thus the ions formed in the ionization chamber are extracted by application of a potential to plate 1 (Fig. 3.1). The half-plates 2 and 2′ displace the beam towards one or the other of them, depending on the difference in potential between them, the aim being to centre the beam on the optical axis. The slits in plates 3 and 5 limit the width and angular dispersion of the beam (i.e. they collimate it), and the half-plates 4 and 4′ maintain the symmetry of the beam. The combined effect of plates 1–5 shown in Fig. 3.1 determines the formation of the ion image at the entrance slit of the analyser. This image is then the 'object' for the analyser.

The source is regulated by these five potentials to produce an intense and stable ion beam, the desired resolution, and an ion yield that is independent of the ion mass.

The electron beam is produced by heating the filament (F) with a current of 1–2 A, and is accelerated by a potential applied between the filament and the electron trap. The electron beam is kept to a diameter of about 1 mm by a weak magnetic field (about 100 gauss) parallel to it and induced by a small permanent magnet. Some electrons may fall on the chamber and are lost. Those traversing the chamber

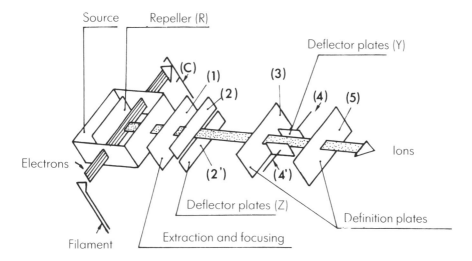

Fig. 3.1 — Ion source.

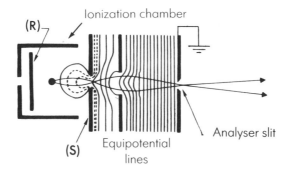

Fig. 3.2 — Ion optics (R = repeller, S = exit slit).

undergo ionizing collisions with the sample molecules, and those which cross the chamber are caught by the collector (or trap, C). The collector current (also called the ionization current) is measured and is proportional to the number of electrons available in the chamber to produce ionization. Under constant conditions the number of ions formed is proportional to this current.

The shape of the electron beam inside the chamber is an important factor since it defines the ionization region and therefore the ion 'object'. The width of this object then determines the ion 'image' and hence is the principal factor affecting the quality of the ion transmission between the source and the analyser.

Ion sources are equipped with a heating system used to volatilize the sample or to clean the source. The temperature of the source is measured by a thermocouple in contact with the chamber. The working pressure in the ionization chamber is usually of the order of $10^{-5}$ torr. The materials used to make an ion source (ionization chamber, plates, etc.) must be non-magnetic to avoid perturbation of the trajectories of the ions and electrons. The chemical activity, catalytic activity and adsorptive power of the chamber walls must be negligible.

The type of spectrum produced when electron-impact ionization is used depends on the energy of the electrons. With low-energy electrons a simple spectrum is produced, but with higher energy many types of fragment ions can be formed in the source, and these are useful in structure determination. Metastable ions with different lifetimes may also be produced (see Chapter 10).

# 4

# Methods for sample introduction

## 4.1 INTRODUCTION

The system used for introduction of the sample into the ion source depends on the physical state of the sample: gas, liquid, solid. The following sections will describe various procedures used for getting the sample, in molecular form, into the ionization chamber.

## 4.2 INTRODUCTION OF A CALIBRATION STANDARD

The reservoir and lines (shown in Fig. 4.1) are pumped down with valve VI open and

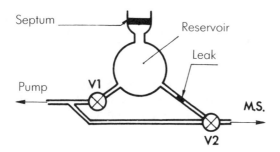

Fig. 4.1 — A simple sample-introduction system for calibration.

V2 closed, until the desired pressure is reached. Vi is then closed and the standard is injected through the septum by syringe. V2 is then switched to connect the reservoir to the mass spectrometer (M.S.), and the substance diffuses through the calibrated leak into the ion source. A fine needle valve may be used instead of the leak. The leak is a small hole (e.g. 2–50 $\mu$m diameter) in a thin (~25 $\mu$m) membrane, such as a gold foil.

### 4.3   INTRODUCTION OF GASES AND VOLATILE LIQUIDS

The inlet system (Fig. 4.2) is pumped down with valves I and II open and III and IV

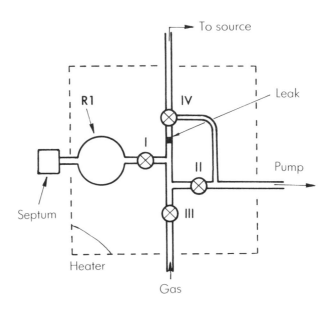

Fig. 4.2 — System for introduction of gaseous or volatile liquid samples.

closed. Valves I and II are then closed and the sample is introduced into the reservoir through the injection port or a septum. Valves I and IV are then opened (II and III are kept closed) and the substance diffuses into the ion source. After the spectrum has been recorded, valve IV is closed. Valves I and II can then be opened and the system pumped down again ready for the next sample. Alternatively, a gas or liquid can be introduced into the reservoir by pumping down as before with I and II open and III and IV closed, then closing II and opening III.

### 4.4   SYSTEM WITH TWO RESERVOIRS

This system is shown in Fig. 4.3. It is operated as follows. Reservoir A can be pumped down with valves III, IV and V open and I, II and VI closed. Reservoir B can be evacuated with valves I, II and IV open and III, V and VI closed. Alternatively, both reservoirs can be pumped down together with valves II, III, IV and V open, and

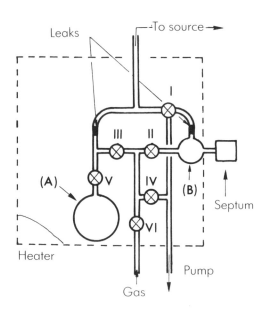

Fig. 4.3 — A double-reservoir introduction system.

I and VI closed. In all three cases all the valves are closed after pumping down is completed.

Suppose that both reservoirs have been evacuated and all the valves closed. Then one substance may be introduced into reservoir B through the septum, and a second substance can be introduced into A through the gas inlet, with valves II, V and VI open (and closed again after introduction of the substance). Alternatively, reservoir A can be fitted with a septum for introduction of a gas or liquid by syringe. Then the sample in B can be introduced into the ion source by opening valve I, or the sample in A can be introduced by opening valve V. A mixture of both can be admitted by opening both I and V. The ratio of the peak intensities is proportional to the relative concentrations. In particular, reservoir A can be used for introduction of a calibration standard. Fine needle valves can be used instead of the calibrated leaks.

## 4.5   INTRODUCTION OF A GAS AT ATMOSPHERIC PRESSURE

The system is shown in Fig. 4.4. Valve V1 is kept permanently open. When V2 is closed, the inlet system is flushed with the sample. When V2 is opened, part of the sample diffuses into the ion source through the leak situated before V2.

## 4.6   DIRECT INTRODUCTION

### 4.6.1   Introduction by a probe

For direct introduction of solid samples, a probe carrying the sample passes into the entrance of the ionization chamber or ion source, through a vacuum lock, as shown

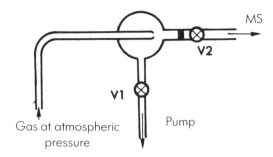

Fig. 4.4 — System for introduction of a sample at atmospheric pressure.

schematically in Fig. 4.5. The method of introduction involves the following steps. The sample holder is cleaned, the sample is placed on it, and the probe is introduced into the airlock, which is then pumped down to the required pressure. The valve between the vacuum lock and the ion source is then opened so that the probe can pass through it to bring the sample into the source, where the sample is desorbed. The spectrum is taken, and the sample holder is withdrawn from the ion source (with automatic closure of the valve) and then from the vacuum lock.

For amorphous or crystalline solids, the substance is placed in a sample holder made of glass or metal, and is desorbed when the probe is heated. The sample holder is finally cleaned.

The desorption of thermally unstable compounds requires rapid heating so that the substance can pass into the gaseous state before thermal decomposition has begun.

If none of the direct introduction methods just described is satisfactory, use of a gold sample holder will often permit a sufficient number of gas-phase molecules to be obtained; the larger the surface over which the sample is spread, the greater the number of molecules vaporized. A filament-type sample holder can also be used for rapid desorption. A drop of sample solution is placed on the filament, the solvent is evaporated, and the filament is introduced into the ion source through the vacuum lock. The sample is then desorbed by heating the filament (by the Joule effect). A 'flash' desorption is obtained. Figure 4.6 shows some typical sample holders.

### 4.6.2  Automatic introduction systems

There are two types of automatic system, which may be called the carousel system and the magazine system. In the first, the samples are loaded on a turntable inside the source housing, and each in turn is brought into position for desorption.

In the magazine system, shown in Fig. 4.7, 20 or more samples can be introduced successively into a direct-probe inlet system. The sample holders contain the substance (a powder or the residue from evaporation of a solution) in a cavity. They are loaded into a guide in the probe, and conveyed in turn by a computer-controlled mechanism to the desorption area, where the sample is desorbed by heating. After desorption the sample holder is transferred into the removal guide and replaced by

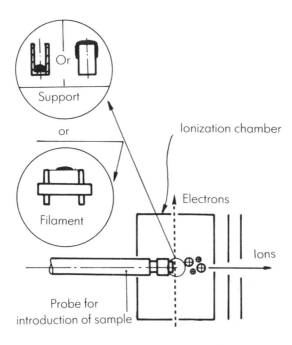

Fig. 4.5 — Principle of direct introduction of sample.

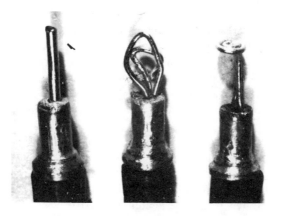

*Figure 4-6*

Fig. 4.6 — Typical sample carriers for direct introduction of sample. (Reproduced by permission, from E. Constantin and R. Hueber, *Org. Mass Spectrom.*, 1982, **17**, 460. Copyright 1982, Heyden Ltd., London.)

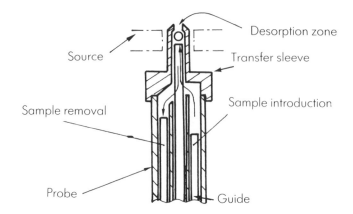

Fig. 4.7 — Automatic probe for sample introduction.

the next sample holder. When the last sample has been desorbed, the probe is withdrawn through the airlock and a new magazine is inserted.

## 4.7 GC/MS AND LC/MS COUPLING

The main problem in coupling a chromatograph to a mass spectrometer is the increase in pressure caused in the ion source by the carrier gas or liquid eluate from the chromatograph.

Two different types of coupling can be used, depending on the type of chromatographic column: with capillary columns direct introduction of the capillary into the ion source is possible, but packed columns require 'separators', a more complex type of interface.

### 4.7.1 Coupling with a gas chromatograph

The molecular jet separator, one of the possible interfaces between a packed column and a mass spectrometer, is based on the fact that the velocity of a gas molecule is a function of the molecular weight (Graham's law of diffusion). Thus low molecular-weight species such as carrier gases or eluent solvents are preferentially removed by pumping down during their passage towards the ion source. The gas flow reaching the source is thereby enriched in the higher molecular-weight species present. In practice, the separators may have one or more stages, each of which can be pumped separately (Fig. 4.8).

A second type of interface, the membrane separator, is based on the relative permeability of a membrane for different molecules (Fig. 4.9).

In the direct coupling interface, the end of the chromatographic capillary passes into the ion source through a glass-lined stainless-steel capillary that is heated to the required temperature by a resistance heater (Fig. 4.10).

### 4.7.2 Coupling with a liquid chromatograph

There are two problems in the analysis of substances eluted from a liquid chromatograph. First, the liquid eluate cannot be easily vaporized, so the GC/MS interfaces

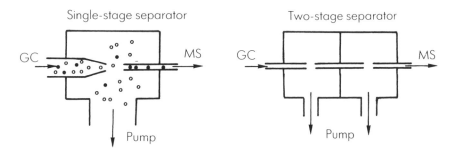

Fig. 4.8 — Molecular jet interface for coupling a gas chromatograph with a mass spectrometer.

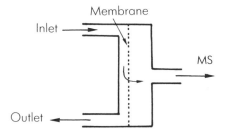

Fig. 4.9 — Membrane interface for coupling a gas chromatograph with a mass spectrometer.

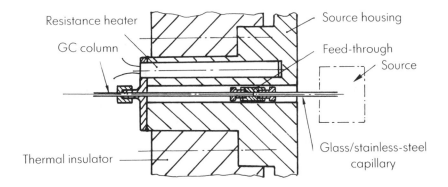

Fig. 4.10 — Direct coupling of a gas chromatograph to a mass spectrometer.

described above in Section 4.7.1 cannot be used, and secondly, the ion-source of the mass spectrometer cannot handle the large amount of gas that would be produced from vaporization of the eluate.

Various approaches have been tried for overcoming these difficulties. Two of these (thermospray and the moving band or wire interface) were mentioned in Section 2.10. The thermospray method will be discussed in more detail in Section 13.2.3. The moving band system is illustrated in Fig. 4.11. The effluent from the

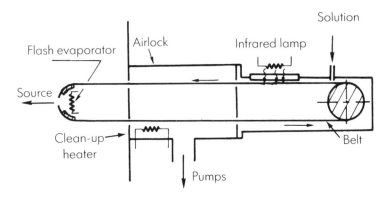

Fig. 4.11 — Schematic diagram of the Finnegan Corporation moving wire interface for coupling a liquid chromatograph to a mass spectrometer. (By courtesy of Finnegan Corporation.)

chromatograph is placed on a continuously moving band (a ribbon or wire) and the solvent is evaporated by infrared heating, followed by desorption of the residue by flash heating in the ionization chamber. A third method utilizes the system already described in Section 4.5, in which the effluent from the chromatograph is vaporized and a fraction is introduced into the mass spectrometer by means of an atmospheric pressure inlet (API) system, which is discussed further in Section 13.2.3.

### 4.7.3  Derivatization

In gas chromatography the compounds to be separated must have sufficient volatility and thermal stability. Similar considerations apply to compounds that are to be examined by mass spectrometry. Compounds having free hydroxyl groups are particularly liable to be thermally unstable and rather involatile, and to increase their stability and volatility sufficiently they are converted into less polar compounds by formation of derivatives by reactions with the polar groups, to prevent the formation of hydrogen bonds. The derivatives commonly used have good chromatographic behaviour and are suitable for GC–MS analysis.

Typical derivatization reactions used for hydroxyl groups include acetylation and trimethylsilylation:

$$ROH \rightarrow ROCOCH_3$$
$$ROH \rightarrow ROSi(CH_3)_3$$

For carboxylic acids and amides, methylation is often used:

$$RCOOH \rightarrow RCOOCH_3$$
$$RCONH_2 \rightarrow RCON(CH_3)_2$$

Methylation has been found to be a good technique for use with nucleosides, and trimethylsilylation and methylation for nucleotides (Figs. 4.12 and 4.13). In addition

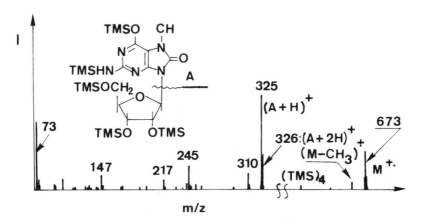

Fig. 4.12.

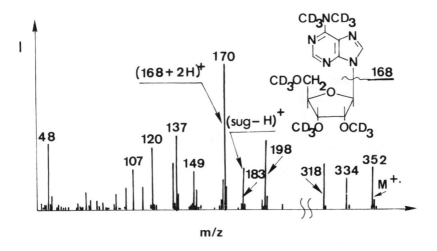

Fig. 4.13.

to trimethylsilylation, trifluoroacetylation and methylation have been used for the nucleic acids. To decrease the polarity of peptides, the $N$-acetyl-$N$, $O$-permethylated or $N$-trifluoroacetyl-$N$, $O$-permethylated derivatives are used. Methylation of peptides results in cleavages which allow identification of the amino-acid sequence of the peptides.

# 5

# The mass analyser

## 5.1 THE MAGNETIC ANALYSER

When an ion (of mass $m$ and charge $z$) leaves the ion source in the mass spectrometer, it has been accelerated by passage through a potential $V$ which imparts to it a kinetic energy $zV$, which is equal to $mv^2/2$, where $v$ is the velocity of the ion. If the ion then enters a magnetic field $B$ perpendicular to the line of flight of the ion, it will be subjected to a magnetic force $F$, given by

$$F = vzB \qquad (5.1)$$

The result is that the ion will follow a circular path with a radius $r$ determined by balancing of the magnetic force $F$ by the centripetal force, which is $mv^2/r$. Thus from

$$mv^2/r = vzB \qquad (5.2)$$

we obtain

$$m/z = rB/v \qquad (5.3)$$

which is the basic equation for application of the magnetic analyser. As can be seen in Fig. 5.1, when ions enter a magnetic field they will follow circular paths of different radii (which depend on the different $m/z$ ratios) and only one of these will be of the correct value for an ion to reach the detector.

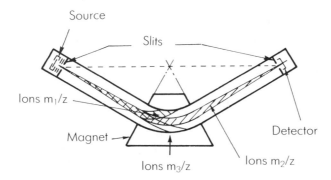

Fig. 5.1 — Principle of the magnetic analyser.

Since $mv^2/2 = zV$, it follows that the velocity of the ion is given by

$$v = \sqrt{2zV/m} \qquad (5.4)$$

and combining this with Eq. (5.3) gives

$$m/z = B^2r^2/2V \qquad (5.5)$$

If $r$ and $V$ are kept constant, $m/z$ is directly related to the magnetic field strength, and ions with different $m/z$ ratios can be passed in turn to the collector by varying $B$ appropriately.

In calculations based on this equation, it is most important to specify the units used, since there have long been mixed systems of units in use in mass spectrometry. In Eq. (5.5), $m$ is expressed in kg, $r$ in metres, $B$ in Tesla ($1\,T = 10^4$ gauss), $z$ in coulombs (C) and $V$ in volts.

In everyday practice in mass spectrometry, it is usual to express $m$ in atomic mass units ($1$ amu $= 1.66 \times 10^{-27}$ kg), $r$ in cm, and $z$ is the number of elementary electronic charges ($e = 1.602 \times 10^{-19}$ C). Thus if $B$ is still expressed in Tesla, Eq. (5.5) becomes

$$m/z = \frac{1.602 \times 10^{-19} \times 10^{-4} B^2 r^2}{2 \times 1.66 \times 10^{-27} V}$$

$$= 4.83 \times 10^3 B^2 r^2/V \qquad (5.6)$$

Thus, for example, for collection of $M^+$ ions ($m/z = 14$), with $V = 2\,kV$ and $r = 24$ cm, the required value of $B$ is

$$B = (14 \times 2000/4.83 \times 10^3 \times 24 \times 24)^{\frac{1}{2}}$$
$$= 0.10 \text{ Tesla}$$

Under the same conditions, for collection of $C_{18}H_{12}S^+$ ions ($m/z = 260$) $B$ would have to be changed to 0.431 Tesla. (With a magnetic field of 0.1 T, $r$ for this ion would be 103.7 cm.)

## 5.2   THE TIME-OF-FLIGHT ANALYSER

In this analyser, ions are produced in the ion source by electron impact, extracted, and introduced into the analyser tube (Fig. 5.2), which is also called the flight tube.

Inside the flight tube the ions move at constant velocity in a straight line, and reach the collector in order of increasing mass, the lightest arriving first. The velocity of the ions on entering the tube is given by Eq. (5.4), so the time ($t$) needed for them to traverse a tube of length $L$ is given by

$$t = L/v = L\sqrt{m/2zV} \qquad (5.7)$$

In this equation the units are $t$ in sec, $L$ in m, $m$ in kg, $z$ in C, and $V$ in volts. Translation into practical units of $t$ in $\mu$sec, $L$ in cm, $m$ in amu, $z$ as number of elementary charges and $V$ in volts, gives

$$t = 10^4 L(1.66 \times 10^{-27} m/2 \times 1.602 \times 10^{-19} zV)^{\frac{1}{2}}$$

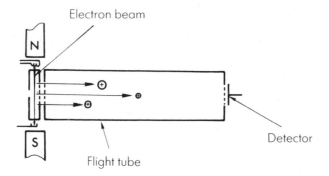

Fig. 5.2 — Principle of the time-of-flight analyser.

$$= 0.72L\sqrt{m/zV}\,\mu\text{sec} \tag{5.8}$$

Thus, for example, with $V = 2\,\text{kV}$ and $L = 100\,\text{cm}$, $t$ for an $H_2^+$ ion would be $2.28\,\mu\text{sec}$, and for a $C_{18}H_{12}S^+$ ion it would be $26.0\,\mu\text{sec}$.

## 5.3 CYCLOTRON RESONANCE ANALYSERS

### 5.3.1 The omegatron resonance analyser

In this analyser, the electrons emitted by heating a filament pass through the ion source chamber and ionize the gas in it. The ions thus formed are acted on by simultaneous application of a radiofrequency electric field $E$ (generated between the plates of a condenser) and a uniform magnetic field $H$ perpendicular to $E$ (Fig. 5.3). If the rf field is in resonance with the cyclotron oscillating frequency of the ions

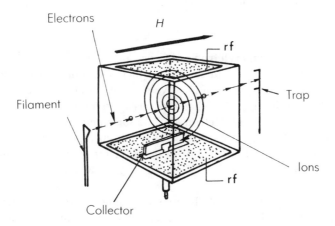

Fig. 5.3 — Principle of the omegatron. (Reproduced with permission, from R. W. Kiser, *Introduction to Mass Spectrometry and its Applications*, p. 72. Copyright 1965, Prentice-Hall, Inc., Englwood Cliffs, NJ.)

in the magnetic field, the ions will follow an Archimedes spiral (of increasing radius). Thus for a given frequency, only one ion species has a path that allows it to reach the detector. By scanning the frequency, it is possible to detect ions with different masses successively. The relation between the mass and the frequency $f$ is given by

$$f = zB/2\pi m \tag{5.9}$$

If $B$ is expressed in Tesla, $m$ in amu and $z$ as number of elementary charges, then

$$f = 1.536 \times 10^7 \, zB/m \, \text{Hz} \tag{5.10}$$

The time $t$ needed for the ion to reach the detector is given by

$$t = 2rB/E \tag{5.11}$$

where $r$ is the distance from the point of origin of the ions to the collector, and $E$ is the electric field strength for the peak signal. If $B$ is expressed in Tesla, $r$ in cm, and $E$ in V/cm, then

$$t = 0.2 \, rB/E \, \text{msec} \tag{5.12}$$

The resolution is given by

$$M/\Delta M = rB^2 z/2Em \tag{5.13}$$

where $M$ is the mass number. With the same units as before for practical measurement, we have

$$M/\Delta M = 4.83 \times 10^3 \, rB^2 z/Em \tag{5.14}$$

For example, for $H^+$ the resonance frequency needed if the magnetic field is 0.4697 T, is 7.21 MHz. If $r$ is 0.7 cm and $E$ is 0.041 V/cm, then $t$ is 1.6 msec, and the resolution is about $1.8 \times 10^4$. For $CO_2^+$, with the same conditions, the frequency would be 164 kHz, $t$ would still be 1.6 msec, but the resolution would be only 413. The resolution can be improved by adjusting the magnetic and/or electric field strengths, and the mass spectrum can be obtained by scanning the magnetic field instead of the frequency.

### 5.3.2  The Fourier-transform cyclotron resonance analyser

This type of mass spectrometer operates on the principle of induction of a current by a charge in a conductor. If a group of positively charged ions moves between two charged plates (A and B in Fig. 5.4), when the ions approach plate B, electrons also accumulate on this plate. This electron flux, determined by the interaction of electrostatic forces, is the 'image' current. This image current produces a potential difference between plate B and earth, and disppears on impact of the ions on plate B. If, however, the ions do not reach plate B but move away from it, the current decreases.

In the Fourier-transform method the ions are made to follow a circular path between the plates by application of a magnetic field, resulting in an alternating image current. The amplitude of this alternating current depends on the number of ions, and the frequency is related to $m/z$ by Eq. (5.9). If several ion species are present, the signal from the amplifier is a composite, with the various frequencies

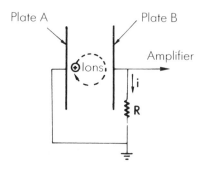

Fig. 5.4 — Principle of the cyclotron resonance Fourier transform spectrometer.

superimposed. This frequency spectrum then corresponds to a mass spectrum, and can be deconvoluted (resolved into its components) by Fourier analysis.

In the Fourier-transform spectrometer, the ions do not reach the collector, but are detected by means of the alternating image current. The first instrument of this type was built in 1981.

## 5.4   THE QUADRUPOLE ANALYSER

A quadrupole filter consists of four parallel electrodes with hyperbolic, elliptical, or circular cross-section (Fig. 5.5). The diagonally opposite electrodes (arranged

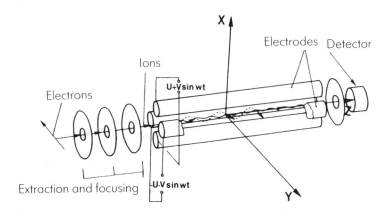

Fig. 5.5 — Principle of the quadrupole spectrometer.

symmetrically with the minimum-radius curve innermost) are at the same potential and are separated by a distance equal to twice the minimum radius of curvature (i.e. $2r_0$). The potential applied to the electrode is of the form

$$\phi_0 = U + V \cos \omega t \qquad (5.15)$$

where $U$ is a d.c. voltage and $V \cos \omega t$ is a radiofrequency potential (frequency $\omega/2\pi$), and the electrodes are coupled so that there is a $180°$ phase difference between the fields applied to the two pairs of rods.

The potential ($\phi$) at any point $(x, y)$ in the quadrupole field is then

$$\phi = \phi_0(x^2 - y^2)/r_0^2 \qquad (5.16)$$

Hence the equipotential lines are equilateral hyperbolas in the $x, y$ plane. The movement of an ion (with mass $m$ and charge $e$)† in the quadrupole filter is expressed by the differential equations

$$m\left(\frac{d^2x}{dt^2}\right) + \frac{2e(U + V \cos \omega t)x}{r_0^2} = 0 \qquad (5.17)$$

$$m\left(\frac{d^2y}{dt^2}\right) - \frac{2e(U + V \cos \omega t)y}{r_0^2} = 0 \qquad (5.18)$$

$$m\left(\frac{d^2z}{dt^2}\right) = 0 \qquad (5.19)$$

These can be transposed into Mathieu equations by substitution of the dimensionless parameters

$$a = 8eU/m\omega^2 r_0^2 \; ; \qquad q = 4eV/m\omega^2 r_0^2 \; ; \qquad \tau = \omega t/2 \qquad (5.20)$$

yielding

$$\frac{d^2x}{d\tau^2} + (a + 2q \cos 2\tau)x = 0 \qquad (5.21)$$

$$\frac{d^2y}{d\tau^2} - (a + 2q \cos 2\tau)y = 0 \qquad (5.22)$$

These equations lead to two types of solution. The first is for stable trajectories, in which the ion oscillates along the $z$ axis, with an amplitude less than $r_0$. The second is for unstable trajectories, in which the amplitude of the oscillations increases exponentially, and the ion eventually hits one of the rods. These trajectories are shown schematically in Fig. 5.6. Obviously the solutions for the $x$ and $y$ directions must both give stable trajectories if the ion is not to be lost by hitting an electrode. (In practice, the actual trajectory for a particular ion depends on the initial position and direction of movement of the ion when it enters the quadrupole, so even 'stable' trajectory ions can be lost; also, the electrodes must be long enough to ensure that the 'unstable' oscillations become large enough).

The conditions for stability can be represented on an $a, q$ diagram, called a stability diagram or a Mathieu diagram (Fig. 5.7). The zone of 'stable' combinations of $a$ and $q$ lies between the two curves representing the limits of stability in the $x$ and $y$

† Here $e$ is used for the charge, to avoid confusion with $z$ used for a rectilinear co-ordinate.

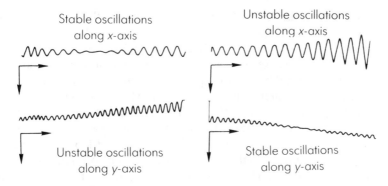

Fig. 5.6 — Stable and unstable trajectories in the quadrupole spectrometer.

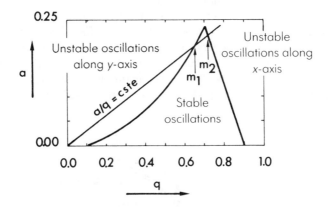

Fig. 5.7 — Stability diagram for the quadrupole spectrometer.

directions. From the dimensionless parameters [Eqs. (5.20)] it is clear that for given values of $U$, $V$, $\omega$ and $r_0$, the $a$ and $q$ values will depend on $m/e$, and since $a/q$ is equal to $2U/V$ and independent of $m/e$, $\omega$ and $r_0$, the $a,q$ co-ordinates corresponding to different $m/e$ values will lie on a line of slope $a/q$ drawn on the stability diagram. This line, called the working line, cuts the stability boundary at two points, corresponding to two $m/e$ values, $m_1/e$ and $m_2/e$. All ions with $m/e$ values beween these will be transmitted by the filter and detected. Ions having the same $m/e$ value will have the same $a,q$ co-ordiantes. The resolution will depend on the slope of the working line. The closer the line comes to the apex of the stability boundary, the greater the resolution. The apex is at $a = 0.237$ and $q = 0.706$. Mass scanning is done by varying $U$ and $V$ simultaneously, while keeping their ratio and $\omega$ constant, or by keeping $U$ and $V$ constant and varying $\upsilon$.

## 5.5  DOUBLE-FOCUSING SPECTROMETERS

These comprise an electrostatic analyser and a magnetic analyser. The term double focusing refers to the two focusing modes: the magnetic analyser offsets angular divergence of the ion beam, giving *directional focusing*, and the electrostatic analyser achieves *velocity* focusing, since ions with different energies but the same mass are brought to a focus. The electrostatic analyser, consisting of two parallel curved plates at a potential difference $E$, exerts a force $F = Ez$ on an ion of charge $z$ and velocity $v$. If this force is just balanced by the centripetal force, $mv^2/r$, the ion follows a circular path with radius $r$. The kinetic energy of the ion when it leaves the ion source is $mv^2/2$, which is equal to $zV$, where $V$ is the accelerating potential. Hence $r = 2V/E$, and all ions with the same kinetic energy follow the same path, irrespective of their mass, and can be brought to a focus. Those with different kinetic energies follow different paths. Thus those with the same $m/z$ value will be brought to different foci according to their velocities. This is velocity focusing. If, after passage through the focal point, this beam then enters a magnetic analyser, it will be slightly divergent, and only ions with the same $m/z$ will be brought to a focus, with as many foci as there are $m/z$ values. This is directional focusing.

The problem is that the locus of the velocity foci is not the same as the locus of the directional foci, but intersects it at a point (the double-focusing point) corresponding to one particular mass. It is necessary to scan the magnetic field to bring ions of different masses to the double-focusing point. By careful design it is possible to reduce the angle between the two focus loci and thus broaden the mass range that can be recorded, as in the Bainbridge–Jordan spectrometer (Fig. 5.8).

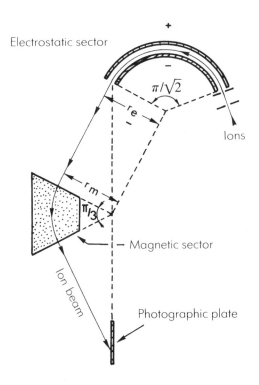

Fig. 5.8 — Bainbridge–Jordan type of spectrometer.

In double-focusing instruments of the Mattauch–Herzog type, the electrostatic field generates an ion beam focused at infinity and dispersed according to energy. The beam is then focused by the magnetic sector, and all ions are simultaneously in focus, irrespective of their $m/z$ values. The advantage of this and the Bainbridge–Jordan type of spectrometer is that a photographic plate can be used for detection. The Nier–Johnson instrument, a development of the Bainbridge–Jordan model, is double focusing at only one point, and uses an electron multiplier for detection, scanning the magnetic field to focus each ion in turn on the detector slit. Both types of instrument give remarkably high resolution.

# 6

# Methods of display, recording and processing of signals

## 6.1 DETECTION

The ions separated by the analyser are detected when they reach the collector. The detection involves production of a signal by a converter that is sensitive to ion impact. The methods used are either photographic or electrical.

### 6.1.1 Photographic detection

This method is markedly less sensitive than electrical detection. Moreover, the response is not a linear function of the energy or number of the ions, or the time of exposure. The method is not suitable for measurement of isotopic abundance or for fragmentation studies, and its sensitivity is low. On the other hand, the photographic plate acts simultaneously as collector, converter and recorder. Other advantages are that it gives the highest possible resolving power for a particular instrument and can give simultaneous recording of the whole spectrum. It is particularly useful in spark source mass spectrometry.

### 6.1.2 Electrical detection

Two main types are used.

#### 6.1.2.1 The Faraday cup

In the *single collector* version, the ion beam for analysis passes through a defining slit (DS in Fig. 6.1) and a guard (or ion-suppressor) slit (IS). These slits eliminate the metastable and scattered ions, and decrease the dispersion of the ion beam. The resolution is thus increased, and hence also the precision of determination of the ion current. The ion suppressor plate is maintained at the source potential and acts as a potential barrier between the analyser and the collector, which can only be passed by ions which have acquired the full (maximum) kinetic energy from the acceleration region. It also repels any secondary ions arising from bombardment of the defining

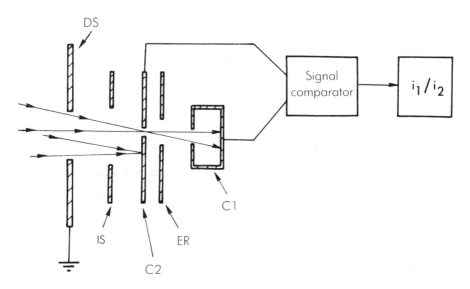

Fig. 6.1 — The dual collector for measuring isotopic abundances. DS = defining slit; IS = ion-suppressor slit; C2 = collector for ion beam No. 2; ER = electron repeller; C1 = Faraday cup collector for ion beam No. 1.

slit. An electron-repeller (electron suppression) plate is often added, to return to the collector any secondary electrons produced by the impact of positive ions on it. An unearthed guard ring is generally placed between the electron suppressor and the collector to prevent any leakage current.

The ion current appearing at the collector is directly proportional to the number of charges per ion and the number of ions. It is independent of the chemical nature and the kinetic energy of the ions and is therefore suitable for measurement of the total ion current. For low currents ($10^{-18}$–$10^{-17}$ A), amplifiers can be connected to the detector. Two types are in general use.

In the *direct current electrometer* a high ohmic resistance is placed between the Faraday cup and earth, to give a voltage output in the mV–V range, which is then impressed on the first grid of the electron amplifier (see Section 6.1.2.2).

In the *alternating current (vibrating reed) electrometer* the d.c. signal is imposed on a vibrating-reed condenser, the output of which is an alternating current, which is then easily amplified.

In the *dual collector* version (Fig. 6.1), one ion beam is collected on a slit plate (C2) and another ion beam by the Faraday cup (C1). The ratio of the two ion-currents ($i_1/i_2$) is found by a null method, in which the current from the amplifier for one collector is cancelled by a fraction of the current from the amplifier for the other collector.

### 6.1.2.2 *Electron multipliers*
The principle on which these are based is the emission of secondary electrons when an energetic particle strikes a suitable surface. The secondary electrons can be

'multiplied' (i.e. increased in number) by being made to strike a second such surface, and so on. There are two main types.

In the *electrostatic multiplier*, the positive ions leaving the analyser are increased in energy by an accelerating voltage and then strike the first plate (the conversion cathode) of the multiplier, and cause the emission of secondary electrons (Fig. 6.2).

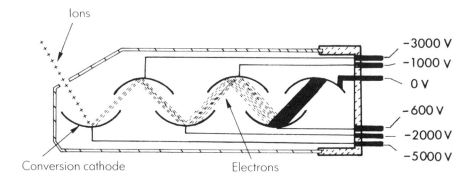

Fig. 6.2 — The electrostatic multiplier. (Reproduced, by permission from R. W. Kiser, *Introduction to Mass Spectrometry and its Applications*, p. 6. Copyright 1965, Prentice-Hall, Inc., Engelwood Cliffs, NJ.)

These electrons are accelerated and directed onto a second electrode (called a dynode) where their impact causes emission of further secondary electrons, which are directed to the next dynode and so on, until the last dynode is reached. The electrons finally fall on the anode (at ground potential). It is thus possible, by use of a sufficient number of conversion stages, to attain a final current of the order of $10^{-9}$ A.

This detector is highly sensitive and gives a rapid response, but is not suitable for the measurement of total current.

The dynodes are usually made of Be-Cu alloy (2% Be), annealed by heating under vacuum or in an inert atmosphere at 200–400°C. This type of multiplier is sensitive to magnetic fields and must be shielded from them.

In the Wiley *magnetic multiplier* (Fig. 6.3), a magnetic field $H$ and an electric field $E$ are used simultaneously, to produce a cycloidal trajectory of the electrons from the conversion cathode. A uniform magnetic field is applied perpendicularly to the optical axis of the ion beam. Instead of dynodes, two strips of semiconductor material (the field strips) are used. Under the influence of the magnetic and electric fields the secondary electrons have a cycloidal path, but strike the field strip again before completing a cycle, producing another burst of electrons, and amplifying the signal.

This multiplier can take various forms, two of which are mentioned here. In the channel electron multiplier ('channeltron') a funnel-shaped hollow tube is used, with the field strips on its inner surface, and the neck may be either straight or curved (the latter to avoid a reverse current due to positive ions accelerated in the opposite direction to the electrons). In the channel electron multiplier array (CEMA, or

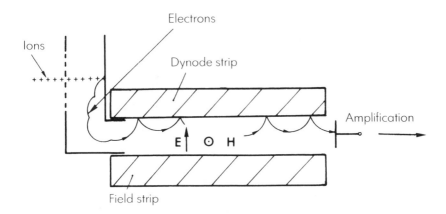

Fig. 6.3 — The Wiley magnetic multiplier.

'channelplate') a large number of very small channels functioning like channeltrons are arranged in parallel. If a radial array is used, the system is particularly useful for studying the angular distribution of ion species resulting from ion-molecule reactions, and for the detection of ions in time-of-flight spectrometers. The ion signal is increased because all the ions arriving on the array at the same time are collected.

## 6.2  RECORDING AND DISPLAY OF RESULTS

### 6.2.1  Galvanometer recording
A mirror galvanometer is used, with the light beam (usually ultraviolet) reflected onto light-sensitive paper. The main advantage of the system is its speed of response (1–100 msec). Several galvanometers, with different sensitivities (e.g. × 1, × 10, × 100), can be grouped and their signals recorded simultaneously. This gives greater precision, because the most appropriate sensitivity can be selected for the measurement.

### 6.2.2  Miscellaneous methods
Recordings can also be made on a chart recorder, or the signals can be displayed on an oscilloscope (and photographed) or stored in a computer.

## 6.3  DATA-ACQUISITION AND PROCESSING

### 6.3.1  Data-acquisition
Microprocessor data-acquisition requires an interface which converts analogue signals into digital form (an analogue/digital converter, ADC) as shown in Fig. 6.4, for storage in the computer memory and for further processing. It should be noted from Fig. 6.4 that though the analogue peak shown is symmetrical, the digitized 'sample' readings ($y_j$) taken at equal intervals will only be symmetrical if one of them coincides with the maximum of the peak. In practice this means that enough

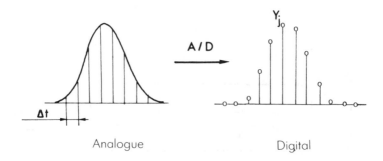

Fig. 6.4 — Analogue to digital signal conversion.

'samples' must be taken to ensure acceptable accuracy, but not so many as to make the process uneconomic. For exponential scanning, the sampling is done at an ADC rate given by

$$S = 0.0023 \, NR/t_{10}$$

where $S$ is the ADC rate (kHz), $N$ the number of samples per peak, $R$ the resolution of the spectrometer, and $t_{10}$ the duration of scan (sec) per decade.

### 6.3.2  Data-processing
The following steps are required.

(1) Filtering of the signal to remove background noise. A digital threshold is selected, and signals lower than this critical value are eliminated electronically or by computer software.
(2) Peak recognition. A program is used which recognizes the beginning and end of a peak, and gives a precise value for its position on the mass scale. To improve the detection of small peaks, a digital smoothing program can be applied before the threshold filtering. One formula that is used for smoothing [1] is

$$y_j = \frac{1}{16} \sum_{i=-2}^{i=+2} c_i \, y_{j+i}$$

with $c_i = 6$ for $i = 0$, 1 for $i = \pm 2$, and 4 for $i = \pm 1$. The value of each successive point is adjusted (smoothed) by use of the two preceding and two following points.
(3) Location of the peak. Various methods can be used for this.

(a)  The centroid of the peak can be determined $(y_c)$; it is given by

$$y_c = \Sigma jy_j / \Sigma y_j$$

where $y_j$ is the digital value for the $j$th sample, taken at time $t_j$ on the scan. The centroid is the weighted mean (or centre of gravity) of the values.

(b) The scan-time $t_m$ corresponding to the maximum of the peak can be determined; if the peak is symmetrical, $t_m = t_c$, where $t_c$ is the scan-time corresponding to the centroid.

(c) The points of inflection of the peak can be calculated and averaged.

(4) Calibration of mass ($m$) and intensity ($I$). Once the positions of the peaks have been established, masses must be attributed to them. To do this, the $y, t$ values for the peaks must be converted into $I, m$ values. The following methods can be used.

If a magnetic analyser has been employed, at constant acceleration voltage, the $m/z$ values can be directly related to the magnetic field strength, which can be measured by a number of techniques, including nuclear magnetic resonance. In many commercial instruments the Hall effect is used to provide a so-called mass marker. An alternative is to operate the analyser at constant magnetic field strength and scan the accelerating voltage. When a quadrupole analyser is used, the $m/z$ can be obtained from the values of $U$ and $V$, $U/V$ and the radiofrequency (see Section 5.4). An alternative is to use known peaks from a reference compound, such as n-octane or perfluorokerosene. Reference peaks selected in the high-mass region of the spectrum are located, and from their mass and appearance time on the mass spectrum, the masses for other peaks can be calculated as follows.

Suppose we have four reference masses $m_1 > m_2 > m_3 > m_4$; then if the scans are from high to low mass the appearance times for these peaks in the scan will be $t_1 < t_2 < t_3 < t_4$. From $t_1$, $t_2$ and $t_3$, the time $t_4$ corresponding to mass $m_4$ can be predicted by using the equation [2]

$$m_4 = \frac{m_3 \exp(t_3 - t_4)}{\tau_{3,4}}$$

where

$$\tau_{3,4} = \tau_{2,3} - (\tau_{1,2} - \tau_{2,3})\,(m_2 - m_4)/(m_1 - m_3)$$

$$\tau_{2,3} = \frac{m_2 \exp(t_2 - t_3)}{m_3}$$

$$\tau_{1,2} = \frac{m_1 \exp(t_1 - t_2)}{m_2}$$

If this predicted time does not agree with the experimental time $t_4$ within a preselected limit, it is substituted for it to obtain a revised value for $\tau_{3,4}$, and this operation is then extended to correct the other $\tau_{i,i+1}$ values. The results can then be used to calculate the mass $m_u$ for an 'unknown' with a peak at time $t_u$ by use of an equation of the form

$$m_u = \frac{m_2 \, (\exp t_2 - t_u)}{\tau_u}$$

where

$$\tau_u = \tau_{2,3} + (\tau_{1,2} - \tau_{2,3}) \, (t_3 - t_u)/t_3 - t_1)$$

To determine the exact mass $(m_2)$ of an unknown peak in a spectrum, from that of a reference peak $(m_1)$, the following equation can be used:

$$m_2 = [\sqrt{m_1} + (t_2 - t_1)/k]^2$$

where $k$ is the slope of a plot of $t$ against $\sqrt{m}$. Various other equations may be used, which generally involve polynomial interpolation.

For dealing with multiplets and metastables, complementary calculation programs are available.

**REFERENCES**

[1] R. A. Hites and K. Biemann, *Anal. Chem.*, 1967, **39**, 965.
[2] W. J. McMurray, in *Mass Spectrometry: Techniques and Applications*, G. W. A. Milne (ed.), Wiley–Interscience, New York, 1971, p. 72.

# 7

# Types of ions, peaks and mass spectra

## 7.1  ION SPECIES

The mass spectrum can contain signals from several species of ions, including the following:

singly charged monatomic ions: $X^+$
singly charged polyatomic ions: $XY \ldots Z^+$
multiply charged ions: $X^{n+}$, $XY \ldots Z^{n+}$
molecular ions: $XY^{+\cdot}$ (formed from the XY molecule; also called parent or parent-molecule ions)
fragment ions: $X^+$, $Y^+$ (formed from the XY molecule)
rearrangement ions: $XRY^+$ (formed from XYR or RXY)
metastable ions: generally formed by decompositions occurring between the ion source and the analyser
secondary ions: formed by reactions between ions and molecules
negative ions: ions (negatively charged) corresponding to those above.

## 7.2  TYPES OF PEAK

The peaks occurring in the mass spectrum correspond to the above-mentioned ion species.

### 7.2.1  The molecular peak

This (also called the parent-molecular peak) corresponds to the ionized molecule $M^{+\cdot}$ (e.g. $CH_4^{+\cdot}$). It is important that it be present in the mass spectrum, as it gives the molecular weight of the substance and therefore enables identification of the compound.

### 7.2.2  Fragmentation peaks

These correspond to the fragment ions (e.g. $CH_3^+$, $CH_2^+$ from methane). They are important for establishing molecular structure.

### 7.2.3  Normal peaks

These are the peaks for all the ions formed in the ion-source, which may be of various types. Figure 7.1 shows the normal peaks found in a mass spectrum of furan.

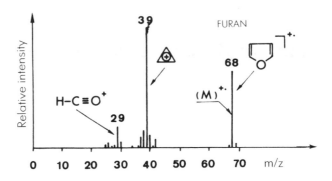

Fig. 7.1 — Mass spectrum of furan, showing the normal peak and fragmentation peaks.

### 7.2.4  Metastable ion peaks

These correspond to the ions formed in metastable transitions (and called metastables or metastable ions), which occur after the ion-beam has left the ion source. They are generally found at non-integral $m/z$ values, and are discussed in detail in Chapter 10.

### 7.2.5  Rearrangement peaks

These are generated by ions in which at least one atom has undergone an interatomic displacement. For example, the ion $CF_2H^+$ may be formed from $CF_3CH_2OH$.

### 7.2.6  Secondary ion peaks

Since these arise from the ions formed in ion–molecule reactions, their intensities will depend on the extent of the corresponding reactions, and hence on the partial pressures of the reactants in the ion source (which is where the reactions take place) and on the residence times in the source. The most important secondary ion peak is the quasi-molecular peak $MH^+$ resulting from transfer of a proton to the molecule, e.g. $X + YH \rightarrow XH^+ + (Y - H)^-$. Methods of ionization that are specific for heavy or unstable substances and those with low volatility, give quasi-molecular peaks, that play the role of the molecular peaks, which are normally absent from these spectra (see also Chapter 11).

### 7.2.7  Isotope peaks

Since many elements have several isotopes, for any given atomic composition the mass spectrum contains peaks corresponding to the different isotope combinations (see also Chapter 17). Thus in the case of methane, the isotope peaks for $^{12}CH_4^+$ ($m/z = 16$), $^{13}CH_4^+$ ($m/z = 17$) and $^{12}CH_3D^+$ ($m/z = 17$) will appear. Note that as the abundance of $^{13}C$ is 1.08% and that of deuterium (D) is only 0.016%, the peak at $m/z = 17$ will be mainly due to $^{13}CH_4^+$.

The isotope peaks corresponding to the natural isotopic composition of the elements play an important role in the identification of a compound. They are also used in following the behaviour of isotopically labelled compounds.

Electron-impact ionization of molecules containing polyisotopic elements such as C, H, N, O, results in peaks at $m/z$ values that are 1, 2, 3, etc. units higher than the peak for the ion containing the most abundant isotope of each element. The relative abundances of the different $m/z$ values can be calculated from the relative abundances of the isotopes concerned, by use of the binomial expansion $(a + b)^n$ where $a$ and $b$ are the natural abundances of the light and heavy isotopes respectively. Thus for a carbon compound, since the abundances are 98.9% for $^{12}C$ and 1.1% for $^{13}C$, the first term of the expansion (i.e. $a^n$) gives the abundance of ions containing only $^{12}C$ atoms, and the second ($na^{n-1}b$) gives the abundance of ions containing one $^{13}C$ atom, which is therefore approximately $1.1n$% (for $n < 10$). When more than one polyisotopic element is present, the product of the individual binomial expansions is used. Hence for a C,H,N compound, the abundance of $^{15}N$, 0.37%, should be taken into account, whereas that of $^2H$, 0.016%, need not. Thus for diaminoethane ($C_2H_8N_2$), to a first approximation the M + 1 peak ($m/z = 61$) will have an abundance of $2 \times 1.1 \times 2 \times 0.37 = 2.9$%. Obviously the calculation becomes more complicated as the number of atoms and hence isotope combinations increases, and details of the calculations have been given by Margrave and Polansky [1]. An increase of 2 units in the $m/z$ value will be obtained for compounds containing oxygen ($^{18}O$ 0.20% abundance), chlorine ($^{37}Cl$ 24.5% abundance), bromine ($^{81}Br$ 49.5% abundance) and so on. Thus for a $C_nH_pN_qO_r$ compound the abundance of the M + 2 peak would be approximately $[(1.1n)^2 + (0.37q)^2 + 0.2r]$%.

### 7.2.8    The base peak
One of the peaks, usually the most intense one, is chosen for normalizing the intensities, and is called the base peak.

### 7.2.9    Multiply charged ion peaks
A species of mass number M can give rise to peaks at $m/z = M$ and $m/z = M/2$, for its singly and doubly charged ions respectively. An example is shown in Fig. 7.2.

### 7.2.10    Multiplets
Ions having the same nominal mass number and charge will produce peaks at the same $m/z$ value, which are not separable by a low-resolution mass spectrometer but may be distinguished if sufficiently high resolving power is available. Examples (for $m/z = 138$) are $C_5NO_4^+$ (exact mass 137.9821), $C_{10}H_4N^+$ (138.0338), $C_{11}H_6^+$ (138.0464), $C_3H_{10}N_2O_4^+$ (138.0635).

Multiplet peaks can be used to determine the exact mass corresponding to an unknown peak, from that for a known multiplet peak.

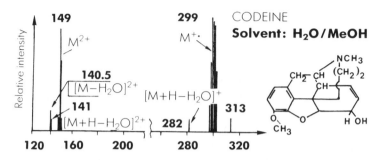

Fig. 7.2 — Mass spectrum of codeine, showing peaks for doubly charged ions. (Adapted, by permission, from H.-J. Schulten, *Int. J. Mass Spectrom. Ion Phys.*, 1979, **32**, 97. Copyright 1979, Elsevier, Amsterdam).

## 7.3  SPECTRA OF THERMALLY UNSTABLE COMPOUNDS

To obtain a molecular or quasi-molecular peak for a thermally unstable substance, a method giving adequate desorption and ionization must be used. The instability of the molecule and the effect of heating are two important parameters in desorption by the Joule effect and ionization by electron impact.

Rapid heating ensures that a sufficient number of molecules will be desorbed for the molecular peak $M^{+\cdot}$ or a quasi-molecular peak $(M+H)^+$ or $(M-H)^-$ to be obtained in the first few spectral scans. Thermal decomposition leads to a progressive decrease in the number of intact molecules and hence to a decrease in the size of the molecular peaks and to their eventual disappearance.

Figure 7.3 shows the evolution of the molecular peak zone for L-arabinose

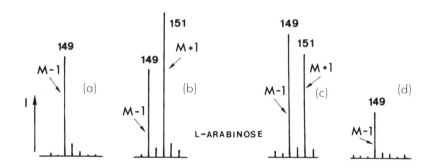

Fig. 7.3 — Changes in mass spectrum of L-arabinose, owing to thermal decomposition. (Reproduced by permission, from E. Constantin, *Org. Mass Spectrom.*, 1983, **18**, 220. Copyright 1983, Heyden Ltd., London.)

$(M = 150)$ on repeated scans. The following features may be observed:

(a)  absence of the molecular peak $M^{+\cdot}$;
(b)  at the start of the desorption the major peak is that for $(M-H)^+$, followed later by the appearance of the peak for $(M+H)^+$;

(c)  both $(M - H)^+$ and $(M + H)^+$ species coexist for a certain length of time, but in changing proportion;

(d)  finally the peak for $(M + H)^+$ disappears first, followed by that for $(M - H)^+$.

The $(M - H)^+$ ions form after electron impact and the loss of a hydrogen atom from $M^{+\cdot}$. During desorption the concentration of M molecules in the ion source increases and ion–molecule reactions take place between the $M^{+\cdot}$ ions and the M molecules, giving $(M + H)^+$ ions, the peak for which will appear when the pressure in the source is high enough to permit reactive collisions. It should also be noted that the decrease in the quasi-molecular peaks for $(M + H)^+$ and $(M - H)^+$ is accompanied by an increase in the peak corresponding to the loss of $H_2O$ from the molecule.

## 7.4   DETERMINATION OF MOLECULAR MASS

Various criteria can be used to identify the molecular peak $M^{+\cdot}$ and/or the quasi-molecular peak $(M + H)^+$.

(1) When a properly chosen desorption and ionization method is used, these two peaks will be those with highest $m/z$ in the spectrum.

(2) When 'soft' ionization techniques are used, such as field or laser desorption, SIMS, fast atom bombardment and chemical ionization, the spectrum is generally very simple, and the $M^{+\cdot}$ and $(M + H)^+$ peaks are sharp and intense. Under certain conditions cationized species may be obtained, with peaks at higher $m/z$ values, as shown later, in Fig. 11.6, for photon impact ionization.

(3) Among the peaks having $m/z$ lower than that for $M^{+\cdot}$, it is easy to identify those for $M - 1$, $M - 15$, $M - 18$, $M - 29$, etc., corresponding to loss of a proton, a methyl group, a water molecule, an ethyl group and so on.

**REFERENCE**
[1] J. L. Margrave and R. B. Polansky, *J. Chem. Ed.*, 1962, **39**, 335.

# 8

# Nominal mass and exact mass

## 8.1 RESOLVING POWER

The resolving power ($R$) of a spectrometer is expressed by

$$R = M/\Delta M \tag{8.1}$$

This ratio represents the ability of a spectrometer to distinguish between ions having neighbouring masses $M$ and $M + \Delta M$.

The resolving power of a magnetic analyser depends on the dispersion of the analyser and on the width of the ion beam reaching the detector. The dispersive power of the analyser is an intrinsic function of its design. The beam width reaching the detector is a function of the exit slit-width of the source ($S_1$) and the entrance slit-width of the detector ($S_2$), the magnification factor of the ion optics system, the velocity dispersion coefficient and the energy inhomogeneity of the ion beam ($\Delta V/V$). For a symmetrical analyser (in which the two slits are equidistant from the analyser and electrical detection is used, with the entrance slit perpendicular to the ion beam), the magnification and the velocity dispersion coefficient can be combined into a single term defined by the radius of deflection of the beam ($r_m$). The resolution is then given by the equation

$$R = \frac{M}{\Delta M} = \frac{1}{(S_1 + S_2)/r_m + \Delta V/V} \tag{8.2}$$

where $V$ is the accelerating voltage and $\Delta V$ is the energy dispersion range for the ions. The contribution of the term $\Delta V/V$ can be reduced by using high energies and an energy filter. If photographic detection is used, $S_2$ equals zero. The energy filter is usually an electrostatic analyser. This creates a double-sector spectrometer.

The energy resolution of the electrostatic sector ($R_e$) is expressed by

$$R_e = r_e/(S_{e1} + S_{e2}) = V/\Delta V \qquad (8.3)$$

where $r_e$ is the sector radius and $S_{e1}$ and $S_{e2}$ are the widths of the entrance and exit slits of the electrostatic sector.

An instrument of this type is called a double-focusing spectrometer (see also Section 5.5) and can separate the closely neighbouring peaks of a multiplet.

It should be noted that all the factors involved in formation of the beam can influence the resolution. Field leaks can therefore cause loss of resolution. Thus space charges arising in the source or analyser can alter the potential distribution, and surface charges may arise from ion bombardment of surfaces. Fringing field effects arise from the lack of well defined boundaries for the magnetic and electric fields, and also affect resolution. For example, for a spectrometer with a magnetic sector having a magnet gap of 1 cm and $r_m = 10$ cm, the resolution could be decreased by about 10%.

A spectrometer can operate at high or low resolution, depending on the width of the slits. The expressions on the right-hand side of Eqs. (8.2) and (8.3) do not contain a mass term, so the resolving power of a magnetic or double-focusing spectrometer is independent of the ion mass. However, the resolution depends on so many factors that its theoretical value is not only difficult to calculate, but may be very different from the practical value, which must be determined experimentally.

## 8.2 EXPERIMENTAL VALUE OF THE RESOLUTION

When an exponential scan is used, the peak width at a given fraction of the peak height is the same for all peaks. Thus if two peaks corresponding to nominal masses $M$ and $M_1$ (see Fig. 8.1) have a baseline width $d$ and are separated by a distance $L$ on

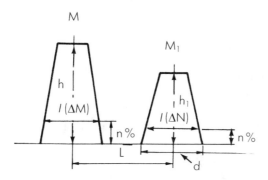

Fig. 8.1 — Definition of resolving power of a spectrometer.

the oscilloscope or recording, they will be completely resolved if $L$ is not less than $d$. For qualitative purposes a certain amount of overlap is acceptable, and the resolution will then depend on the peak width at some fraction ($n\%$) of the peak height. Thus if the '$n\%$ peak width' is $l$, the height of the 'valley' floor when the two peaks overlap will be $n(h + h_1)/100$, where $h$ and $h_1$ are the peak heights. A commonly used definition of the resolution is the height of the valley floor between two peaks of equal height, and a '10% valley' corresponds to overlap of the peaks at 5% of peak height. Thus the resolution is then defined by

$$R = M/\Delta M = ML/l(M_1 - M) \tag{8.4}$$

since $l(M_1 - M)/L$ corresponds to the mass difference ($\Delta M$) represented by the peak width at $n\%$ (in this case 5%) of peak height.

It follows that although the resolution is independent of the value of $M$, the ability to distinguish between $M$ and $M + 1$ is not. For example, a resolution of 100 would allow differentiation between two peaks corresponding to $M = 100$ and 101, but not between the peaks for $M = 300$ and 301. The resolution therefore has to be adjusted by changing the parameters in Eqs. (8.2) and (8.3), e.g. by adjusting the slit-widths.

The effect of the peak heights (when they are very different) on the height of the valley floor will obviously be of importance in quantitative work and in detection of trace species.

## 8.3  NOMINAL MASS

In the mass spectrum a peak located at a nominal mass number $M$ can arise from various ion species. For example, the peak at mass number 28 can be due to ion species such as $CO^+$, $N_2^+$, $CH_2N^+$, $C_2H_4^+$, and a peak at mass number 249 could come from $C_{20}H_9^+$, $C_{19}H_7N^+$, $C_{19}H_5O^+$, $C_{19}H_{21}^+$, $C_{13}H_{19}N_3O_2^+$ etc. when these species are composed of the most abundant isotopes.

## 8.4  EXACT MASS

The masses of the species listed above are not exactly equal, however, and differ by up to almost 0.1%. The exact masses for $^{12}C$, $^1H$, $^{14}N$ and $^{16}O$ on the carbon-12 scale are:

$$^{12}C = 12.0000000$$
$$^1H = 1.0078246$$
$$^{14}N = 14.0030738$$
$$^{16}O = 15.9949141$$

so those for the ions corresponding to mass number 249 are:

| | |
|---|---|
| $C_{20}H_9^+$ | 249.0704 |
| $C_{19}H_7N^+$ | 249.0580 |
| $C_{19}H_5O^+$ | 249.0341 |
| $C_{19}H_{21}^+$ | 249.1643 |
| $C_{13}H_{19}N_3O_2^+$ | 249.1479 |

Note that by convention the mass of the electron (0.0005486 amu) is not deducted. The masses have been rounded off.

## 8.5   RESOLUTION NEEDED TO SEPARATE A DOUBLET

The exact mass of $C_{19}H_7N^+$ is 249.0580 and that of $C_{19}H_5O^+$ is 249.0341. This doublet could be separated by use of a spectrometer having a resolving power of $249/(249.0580 - 249.0341) \sim 1.042 \times 10^4$.

## 8.6   DETERMINATION OF EXACT MASS BY THE PEAK-MATCHING METHOD

As the name implies, this method of finding the exact mass $M_x$ of an unknown peak makes use of a reference peak for a known exact mass $M_0$.

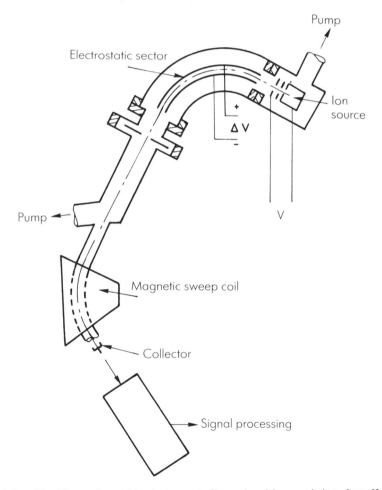

Fig. 8.2 — The Nier peak-matching instrument. (Reproduced by permission, from K. S. Quisenberry, T. T. Scholman and A. O. Nier, *Phys. Rev.*, 1956, **102**, 1071. Copyright 1956, American Institute of Physics, New York.)

In the normal working mode of a spectrometer, the singly charged ions formed in the ion source and then accelerated receive the same kinetic energy, $eV$. Therefore when the two ions are subjected to the same magnetic field, their trajectories will be slightly different and two distinct peaks, separated by a distance $L$, will be seen on the oscilloscope screen.

Peak matching is based on the fact that in the same magnetic field $B$, ions with masses $M_x$ and $M_0$ will follow the same trajectory if their kinetic energies are in the ratio $eV_0/eV_x = M_x/M_0$, where $V_0$ and $V_x$ are the acceleration voltages for the $M_0^+$ and $M_x^+$ ions respectively. Peak matching then consists of varying the kinetic energy of the $M_x^+$ ions from $eV_0$ to $e(V_0 + \Delta V)$ until $M_x^+$ follows the same trajectory as the reference ion $M_0^+$ accelerated by voltage $V_0$. This is done by cycling the acceleration voltage over the range $V_0 \pm \Delta V$ until both peaks appear at the same position on the screen. The persistence time of the signal on the screen allows simultaneous display of the two peaks and variation of $\Delta V$ until the peaks are superimposed. The ratio $V_0/V_x$ is measured with high precision by use of calibrated resistance boxes. To maintain the precision the ratio of the masses should lie between 0.9 and 1.1.

The instrument designed by Nier is shown schematically in Fig. 8.2.

## 8.7  USE OF HIGH RESOLUTION

High resolution has two main uses. First, to determine the exact mass of an ion species; a special application is to an ion-molecule derived from an unknown substance, since knowledge of its exact mass allows its atomic composition to be established. Secondly, the exact mass provides evidence for the presence of a particular ion species in a mixture.

# 9

# Analysis of carbon isotopes

## 9.1 INTRODUCTION

Mass spectrometry is an indispensable tool for the determination of isotopic abundances. Two types of isotope are concerned — stable and radioactive. There are also two types of sample: those with 'natural' isotopic composition, and those in which the concentration of one or more isotopes has been deliberately increased or decreased relative to the natural value.

### 9.1.1. Stable isotopes

The use of stable isotopes in the study of various processes is important, as it avoids all the inconveniences that can be associated with the use of radioactive isotopes. Stable isotope measurements are used for two main purposes: the determination of natural abundances, and the study of processes by using isotopic labelling or a standard of known concentration differing from the natural concentration.

Isotopic labelling mainly involves deuterium ($^2$H, or D), but $^{13}$C, $^{18}$O and $^{15}$N are also used as labels, and the products act as internal standards in an analysis. An example is the use of deuterated morphine to determine the amount of morphine present in the blood or brain. Another is the study of the effect or metabolic pathway of drugs, for example, monitoring the concentration of deuterated disopyramide [1] or of disopyramide phosphate [2] labelled with $^{13}$C or $^{15}$N.

Isotopes of Ca, Zn, Cu, Fe and Se have been used to obtain information on the assimilation (and its mechanism) of these elements in different environments, e.g. determination of trace metals in body tissues and fluids.

## 9.2 DETERMINATION AND APPLICATIONS OF THE $^{13}$C/$^{12}$C RATIO [3]

Studies based on the isotopic composition of carbon began soon after the discovery of $^{13}$C. Measurements of the $^{13}$C/$^{12}$C ratio revealed that its value depends on the origin of the carbon. The measurements require high precision, since $^{13}$C represents only 1% of the total carbon present, and the maximum relative variation in the $^{13}C/^{12}C$ ratio is only about 5%. The method requires a spectrometer with several collectors, and standards with known $^{13}C/^{12}C$ ratios.

Particular fields of application include archaeology, geochemistry, food chemistry and medicine.

### 9.2.1   The $\delta^{13}C$ factor

Variations in the abundance of $^{13}C$ are very slight. They are defined in the form of a factor $\delta^{13}C$, defined by

$$\delta^{13}C = 1000 \left[ \frac{(^{13}C/^{12}C)_{sample}}{(^{13}C/^{12}C)_{std}} - 1 \right] \%o \qquad (9.1)$$

It is therefore the relative difference (in parts per thousand) between the $^{13}C/^{12}C$ ratios for the sample and the standard, referred to the standard. Various standards have been established for determination of $\delta^{13}C$. The classical standard for $^{13}C/^{12}C$ is calcium carbonate obtained from Southern Californian marine fossils (*Belemnitella Americana*), which has a very high concentration of $^{13}C$, so the $\delta^{13}C$ values obtained by its use are generally negative. Secondary standards are also used, and laboratories specializing in this type of work often make their own.

Equations analogous to (9.1) are used to define variations in the $^{15}N/^{14}N$, $^{18}O/^{16}O$ and $^{34}S/^{32}S$ isotopic ratios.

The carbon in the sample for determination of $\delta^{13}C$ for $^{13}C/^{12}C$ is converted into carbon dioxide, which is then analysed by mass spectrometry. Equation (9.1) is then applied to the ratios of the signal intensities at $m/z$ 45 (for $^{13}C^{16}O_2$) and 44 (for $^{12}C^{16}O_2$) for the sample and the standard. If the conditions are chosen so that the intensities of the $m/z$ 44 peaks are identical, then Eq. (9.1) is simplified to

$$\delta^{13}C = 1000 \left[ \frac{^{13}C_{sample}}{^{13}C_{std}} - 1 \right] \%o \qquad (9.2)$$

or in terms of the intensities ($I$) for the peaks at $m/z$ 45,

$$\delta^{13}C = 1000 \left[ \frac{I_{sample}}{I_{std}} - 1 \right] \%o \qquad (9.3)$$

The precision of the $\delta^{13}C$ value is better than $\pm 0.5\%o$.

### 9.2.2   Variations of $^{13}C/^{12}C$ in the environment

Figure 9.1 shows some typical ranges of the $\delta^{13}C$ values found for carbon in various naturally occurring systems. The following observations may be made.

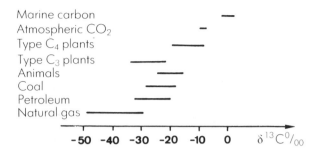

Fig. 9.1 — $\delta^{13}$C‰ ranges for various types of carbon compounds in the environment. (Adapted, by permission, from H. W. Krueger and R. H. Reesman, in *Mass Spectron. Rev.*, 1982, **1**, 205. Copyright 1982, John Wiley & Sons Inc., New York.)

For land plants the $\delta^{13}$C value varies from $-8$ to $-35$‰, depending on the photosynthesis mechanism involved. The type $C_4$ plants fix atmospheric carbon dioxide as malic and aspartic acids (which contain four carbon atoms in the molecule, hence type $C_4$), whereas the type $C_3$ plants fix it by forming phosphoglyceric acid. The $\delta^{13}$C values for mineral oils vary from $-18$ to $-13$‰, which corresponds to the range for the $C_3$ type plants. For animals the values are between $-15$ and $25$‰; various types of sample have been analysed, such as hair and muscle.

### 9.2.3  Use of $\delta^{13}$C to detect adulteration of food [3]

The usual methods of analysis sometimes prove limited in the detection of certain types of fraud, such as adulteration by substitution of components. Analysis by means of $^{13}$C analysis is conclusive, however, provided that the genuine component has a $\delta^{13}$C value sufficiently different from that of the adulterant.

An example is the testing of honey and apple juice in the United States, for the addition of high fructose corn syrup (which is obtained from maize and is rich in fructose). Corn syrup is cheaper than cane sugar, and has been introduced in the food industry as a substitute for cane sugar in various products, particularly in honey.

Honey is made by bees from the pollen of plants in which the photosynthesis mechanism is of the $C_3$ type, whereas corn employs the $C_4$ type. As shown in Fig. 9.1, the $\delta^{13}$C values for the two types of plant are quite different, and so are those for high fructose corn syrup and the fructose in pure honey. Figure 9.2 shows histograms of the $\delta^{13}$C values found for a number of samples of pure honey, and for high fructose corn syrup. For an adulterated honey the $\delta^{13}$C value will be the average for the components concerned. Figure 9.3 shows the values found for samples of commercial honey — the hatched portion of the distribution indicates samples found to be pure honey, and the remainder of the samples were adulterated with corn syrup. It may be noted that because the ranges for $C_4$ and $C_3$ type plants have only a small gap between them, for honey samples with a $\delta^{13}$C value in the range from $-23.5$ to $-21.5$‰ there is a slight risk of false classification, and a confirmatory test is recommended.

The same method has been used successfully to test fruit juice, and Figs. 9.4 and

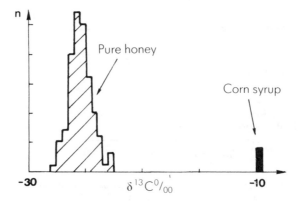

Fig. 9.2 — Carbon isotope analyses of pure honey and corn syrup. (Adapted, by permission, from H. W. Krueger and R. H. Reesman, in *Mass Spectrom. Rev.*, 1982, **1**, 205. Copyright 1982, John Wiley & Sons Inc., New York.)

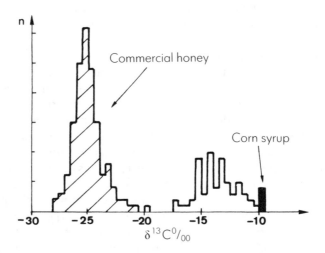

Fig. 9.3 — Carbon isotope analyses of commercial honeys (1980–81). (Adapted, by permission, from H. W. Krueger and R. H. Reesman, in *Mass Spectrom. Rev.*, 1982, **1**, 205. Copyright 1982, John Wiley & Sons Inc., New York.)

9.5 show the values for pure apple juice and for some commercial samples (many of which were shown to be adulterated). The 'risk region', in which adulteration is not conclusively established, is obviously much wider in this case.

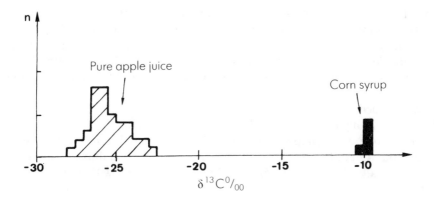

Fig. 9.4 — Carbon isotope analyses of pure apple juice and corn syrup. (Adapted, by permission, from H. W. Krueger and R. H. Reesman, in *Mass Spectrom. Rev.*, 1982, **1**, 205. Copyright 1982, John Wiley & Sons Inc., New York.)

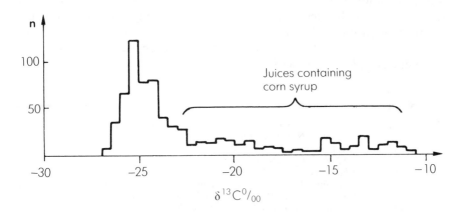

Fig. 9.5 — Carbon isotope analyses of commercial apple juice or concentrate. (Adapted, by permission, from H. W. Krueger and R. H. Reesman, in *Mass Spectrom. Rev.*, 1982, **1**, 205. Copyright 1982, John Wiley & Sons Inc., New York.)

## 9.3  CARBON-14 DATING

### 9.3.1  Introduction

The classical methods of dating by mass spectrometry establish to within about 80 years the dates of events which occurred less than about 40000 years ago. This time limit is due to the sensitivity of the method. The concentration of $^{14}C$ in samples older than this is too low for it to be detected by standard mass spectrometers. These methods also require a sample containing several grams of carbon.

The period between 40000 and 75000 years ago is an important one for archaeologists and anthropologists, however. A more elaborate technique is used for this type of sample, based on use of a tandem particle accelerator coupled with a mass spectrometer to obtain a pure beam of $^{14}C^-$ ions [2]. This method is highly sensitive and requires only a few mg of carbon; hence the interest in its use for measuring concentrations of $^{14}C$ lower than those detectable by the classical methods.

An important recent application was the dating of the Shroud of Turin [5,6].

### 9.3.2 Method

First a sample of pure graphitic carbon is prepared from the object to be dated. In the case of a bone, for example, hydroxyproline is extracted from the collagen of the bone and purified, then converted successively into carbon dioxide, acetylene and carbon. Finally $^{14}C^-$ ions are produced by bombardment with a beam of caesium ions:

$$^{14}C + Cs^+ \rightarrow {}^{14}C^-$$

The big advantage of doing this is that no $^{14}N^-$ ions are produced, because this species is unstable. These would have the same nominal mass as the $^{14}C^-$ ions. If a positive ionization method were used, both $^{14}C^+$ and $^{14}N^+$ ions could be formed. The $^{14}C^-$ ions are directed to the positively charged high-voltage terminal of the tandem accelerator ( + 2.5 MV), which strips off several electrons from the anion, leading to $^{14}C^+$, $^{14}C^{2+}$, $^{14}C^{3+}$ etc. The beam is then magnetically filtered so that only a mono-energetic $^{14}C^{n+}$ species is counted.

**REFERENCES**

[1] N. J. Haskins, G. C. Ford, R. F. Palmer and K. A. Waddell, *Biomed. Mass Spectrom.*, 1980, **7**, 74.
[2] N. J. Haskins, G. C. Ford, P. N. Ford and K. A. Waddell, *Advan. Mass Spectrom.*, 1980, **8**, 286.
[3] H. W. Krueger and R. H. Reesman, *Mass Spectrom. Rev.*, 1982, **1**, 205.
[4] R. E. M. Hedges and J. A. J. Gowlett, *Sci. American*, 1986, **254**, No. 1, 82.
[5] M. Warner, *Anal. Chem.*, 1989, **61**, 101A.
[6] P. E. Damon, D. J. Donahue, B. H. Gore, A. L. Hatheway, A. J. T. Jull, T. W. Linick, P. J. Sercel, L. J. Toolin, C. R. Bronk, E. T. Hall, R. E. M. Edges, R. Housely, I. A. Law, C. Perry, G. Bonani, S. Trumbore, W. Woelfli, J. C. Ambers, S. G. E. Bowman, M. N. Leese and M. S. Tite, *Nature*, 1989, **337**, 611.

# 10

# Metastable ions

## 10.1 LIFETIME OF IONS AND CONCEPT OF METASTABILITY

The ions formed in the ion source have certain lifetimes, depending on their internal energy. Unless it is sufficiently stable to reach the detector without decomposition, an ion formed in the ion source can decompose either before or after it leaves it. If it decomposes in the ion source, then the fragments may reach the detector undecomposed, or may themselves decompose on the way there. This decomposition is a statistical phenomenon and the ratio of the number of ions ($N$) reaching the detector to the number ($N_0$) formed in the source is a function of the lifetime of the species, and of the time taken to reach the detector. Typical residence times in the various parts of a double-focusing spectrometer (8 kV acceleration voltage, $m/z$ 100) are 1–5 $\mu$sec in the source, 3–4 in the first field-free region, 2–3 in the magnetic analyser, 7–9 in the second field-free region, 4–5 in the electrostatic analyser, and 1–6 in the third field-free region, before entry to the collector. It follows that ions with a rate constant greater than $10^6$ sec$^{-1}$ for their decomposition will undergo reaction in the source and the daughter ions will give appropriate normal peaks. If the rate constant if sufficiently low (say $10^4$ sec$^{-1}$) the parent ion will be collected without fragmentation. Ions with decomposition rate constants intermediate between these can decompose during passage through the spectrometer, and these are called metastable ions (or simply 'metastables'), and undergo metastable transitions to give daughter ions that yield so-called metastable peaks, which are diffuse peaks of low intensity in the 'normal' mass spectrum and usually occur at non-integral $m/z$ values.

The characteristics of the metastable peaks depend on the type of instrument and the region in which the transitions take place. If the daughter ions are produced in the acceleration region in the ion source in a single-focusing instrument their kinetic energy will vary according to the field experienced by the parent ion before its decomposition; as this can vary from zero to the maximum potential, the result is a contribution to the background continuum. If they are formed within the magnetic analyser they should be lost to the walls of the analyser and hence not be detected.

The situation for decomposition in the first field-free region is discussed in detail below. Ions formed in the field-free region before the collector will be collected at the same $m/z$ value as the parent ions.

For double-focusing instruments, the characteristics of the metastables depend on the geometry of the instrument. With the normal geometry instrument (electro-static analyser first), daughters formed in the acceleration region are not transmitted by the electric sector. Ions formed within the sector are not normally transmitted. Ions formed in the second field-free region, between the two analysers, behave in the same way as those formed in the first field-free region in a single-focusing instrument. With inverse geometry (magnetic sector first), no metastable peaks appear in the spectrum.

In a time-of-flight spectrometer, if no translational energy change takes place in the fragmentation process, all the particles produced will continue to travel at the same speed, and arrive at the detector simultaneously. They can be distinguished by application of a retarding potential to a grid placed just before the detector. The order of arrival will then be neutrals first, parent ions second, daughter ions last.

## 10.2   IMPORTANCE OF STUDY OF DECOMPOSITION OF METASTABLE IONS

Study of metastable ions is very important because it establishes the relationship between parent ion and daughter ion, and hence the nature of the decomposition reaction. Decompositions taking place after the ion source are generally the same as those occurring within it, so the fragmentation pattern established by studying the metastable ions corresponds to that taking place in the ion source.

## 10.3   METHODS FOR DETECTING METASTABLE DECOMPOSITIONS

Various methods have been introduced for detection of metastable ions. They have the following aims.

(a) To show decompositions taking place in different field-free regions of the spectrometer and to determine the mass of the daughter ions.
(b) To establish the parent–daughter relationships with a view to finding all the ions formed from a particular parent (see Sections 10.3.1–10.3.3) or all the parent species that give the same daughter (see Sections 10.3.1 and 10.3.4).

All techniques for studying metastable ions are based on the fact that these ions have a kinetic energy that is different from that of the ions formed in the source. Detection of these ions therefore requires a change in the values of the electric, magnetic and acceleration fields so that the ions can be transmitted through the analysing sectors.

The basic relations and characteristics of the different methods used are outlined below.

### 10.3.1   The defocusing method

As shown in Section 5.1, an ion of mass $M$, and charge $z$, formed in the source and accelerated by a voltage $V$, will have a kinetic energy $V$ (which is equal to $Mv^2/2$) and hence a velocity $v = \sqrt{2zV/M}$. Suppose that the ion is decomposed after leaving the acceleration zone, to form an ion of mass $m$. Then the daughter ions will have a kinetic energy that is a fraction $m/M$ of that of the parent ions, but will also have the same velocity as the parent ions.

The magnetic field strength $B_M$ required for the trajectory of the parent ions to have the radius $r$ needed for them to reach the detector is

$$B_M = Mv/zr$$

and substitution for $v$ gives

$$B_M = \sqrt{2MV/zr^2}$$

The corresponding value for the daughter ions is given by

$$B_m = mv/zr$$

and substitution for $v$ gives

$$B_m = \frac{m\sqrt{2zV/M}}{zr} = \sqrt{2m^2V/Mzr^2}$$

so when a magnetic analyser is used, the peak for the daughter ion appears when the magnetic field strength is that corresponding to detection of an ion with mass $m^* = m^2/M$, and is positioned at $m/z = m^*$ in the spectrum. For example, if $M = 264$ and $m = 69$, $m^* = \sqrt{69 \times 69/264} = 18.04$.

The defocusing technique [1] consists of displacing the peaks in the spectrum until the peak for the daughter ion $m$ appears at the position corresponding to the nominal mass $M$ of the parent ion. This is done by setting the magnetic field so that only the ions with $m/z = M$ are collected when the accelerating potential is $V_0$, then increasing this potential to $V_1 = MV_0/m$ so that the kinetic energy of the daughter ions formed in the metastable transition is just equal to that needed for them to pass through the magnetic sector and be collected at the exit slit, since

$$B_M = \sqrt{2MV_0/zr^2} = \sqrt{2mV_1/zr^2}$$

The accelerating potential is scanned until the peak for $M$ is replaced by that for $m$. The value of $m$ is then calculated from $m = MV_0/V_1$. In this way all the metastable transitions arising from a given parent ion can be determined, since each in turn will be defocused.

Alternatively, if a given metastable transition is selected and the accelerating potential set so that the peak appears at a position corresponding to the nominal mass, then *decreasing* the accelerating voltage will displace the metastable peak in the opposite direction and result in its replacement by the peaks for the parent species that can give rise to it. In both cases the accelerating voltage scale becomes in effect a mass scale, since $m_i V_i$ is a constant. To keep the accelerating voltage within reasonable limits, it is best to start at a value in the range 2–4 kV.

In general, it is better to use a double-focusing instrument with normal geometry, keep both the magnetic field and ion-source accelerating voltage constant and scan the field potential of the electrostatic sector. If $V_E$ is the electrostatic sector potential necessary for the parent ions with mass $M$ to arrive at the detector, then the potential $V_E'$ needed for transmission of the metastable ion with mass $m$ is given by

$$V_E' = mV_E/M$$

and the metastable peak will appear in the spectrum at the apparent mass $m^*$.

The procedure for use of the defocusing method with a double-focusing instrument with normal geometry depends on the region in which the metastable transition takes place. If it is the first field-free region (i.e. before the electrostatic sector), the accelerating voltage is scanned and the electrostatic sector potential is kept constant, with the magnetic sector tuned so that ions with mass $m$, *formed in the source*, would arrive at the detector.

### 10.3.2  The MIKES technique [2–6]
In this (the <u>m</u>ass-analysed <u>i</u>on <u>k</u>inetic <u>e</u>nergy <u>s</u>pectrometer technique) a double-focusing instrument with reverse geometry is used, with a collision chamber placed between the magnetic sector and the electrostatic sector (Fig. 10.1), and the following sequence of events takes place.

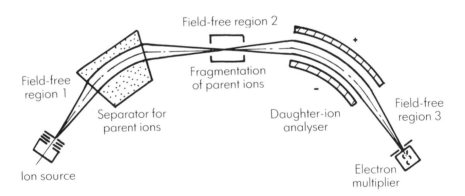

Fig. 10.1 — Principle of MIKES.

1. Ions with different masses ($M_i$) leave the ion source and are accelerated in the acceleration region.

2. From this mixture of primary ions, the species with mass $M$ is separated by the magnetic field, and has velocity $v$.

3. The chosen ions enter the collision chamber, where fragmentation results in production of daughter ions with masses $m_1$, $m_2$, etc., and the same velocity as the parent ion, $v$. The fragmentation may be induced by collisions with molecules of an inert collision gas introduced into the chamber.

4. The parent ions $M$ will traverse the electrostatic sector when its applied potential is $E_M$, and the daughter ions will traverse it at applied potentials of $E_1$, $E_2$ etc. Since the force exerted by a potential $E$ on an ion with charge $z$ is $Ez$ and this is just equal to $mv^2/r$ where $m$ is the mass and $v$ the velocity of the ion, and $r$ the radius of its path through the analyser, it follows that for ions with equal $z$ and equal $v$,

$$M/E_M = m_1/E_1 = m_2/E_2 \ldots$$

so the daughter masses are proportional to the potentials at which they are detected, and the potential scale for the electrostatic sector corresponds to a mass scale.

Thus for a chosen parent ion $M$, the masses of the ion fragments obtained from it can be determined. In the absence of a collision gas, the spontaneous fragmentation of the parent ion in the second field-free region is detected. Introduction of a gas into the collision chamber increases the peak intensities, and in certain cases may lead to formation of new product ions.

This technique was also known as DADI (direct analysis of daughter ions). It allows identification of all the products from a selected ion.

### 10.3.3   The *B/E* method [6]

This method involves simultaneous scanning of the magnetic and electrostatic fields in a double focusing spectrometer.

The primary beam from the source contains ions of mass $M$. Some of these can undergo fragmentation in the field-free region between the two sectors, and the daughter ions, mass $m_1$, will have the same velocity as their parent. The fragmentation may be spontaneous or induced by a collision gas as in the MIKES technique (Section 10.3.2). The mass/voltage ratio will be a constant (also as in the MIKES method). For the parent and daughter ions to traverse the magnetic sector in succession, the magnetic field strength must be matched to their momenta, $m_i v_i$. Since all the velocities are equal and the radius of the trajectory through the magnetic sector is fixed, then from Eq. (5.3)

$$M/B = m_1/B_1 = m_2/B_2 \ldots$$

Hence for a particular ion to reach the detector, the following conditions are necessary.

| Mass | Electric field | Magnetic field |
|------|----------------|----------------|
| $M$ | $E$ | $B$ |
| $M_1$ | $E_1$ | $B_1$ |
| $m_2$ | $E_2$ | $B_2$ |

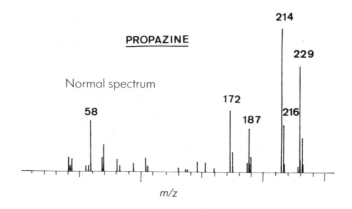

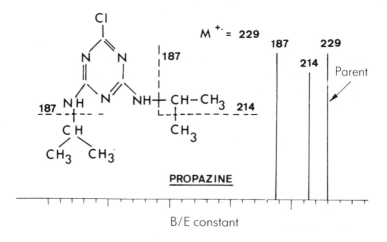

Fig. 10.2 — Mass spectrum obtained by the *B/E* technique.

Since $M/m_1 = B/B_1 = E/E_1$ etc., it follows that $B/E = B_1/E_1$ ... and that this ratio is a constant. Hence if $B$ and $E$ are first set so that $M$ is detected, then changing both values simultaneously so that their ratio remains constant will allow successive detection of the daughter ions $m_i$ and hence identification of all the species produced by fragmentation of a particular parent species.

### 10.3.4   The $B^2/E$ method [6]

This method is based on the following considerations. The momentum of $m_2^+$ ions formed in the ion source with an acceleration voltage $V$ is $\sqrt{2zVm_2}$ and the translational energy is $zV$. If the electric sector voltage and magnetic field are set at $E$ and $B$ respectively, so that the $m_2^+$ ions formed in the ion source are transmitted, then any $m_2^+$ ions formed by a fragmentation of parent ions $m_1^+$ in the first field-free region would not be transmitted, since their momentum would be $m_2\sqrt{2zVm_1}/m_1$, and translational energy $m_2 z V/m_1$. If $V$ is kept constant, then to transmit the product $m_2^+$ ions through the electric and magnetic sectors it would be necessary to use an electrostatic sector voltage $E' = m_2 E/m_1$ and magnetic field $B' = B\sqrt{m_2/m_1}$. Thus if $E$ and $B$ are varied simultaneously, with $B^2/E$ kept constant, it is possible to identify all parent ions that can give rise to the same product ion.

### 10.3.5 Applications
The scope of application of metastables for analytical purposes and gaining structural information is remarkably wide, and has been reviewed in specialist reports, e.g. [7], and elsewhere [4,8].

**REFERENCES**

[1] T. W. Shannon, T. E. Mead, C. G. Warner and F. W. McLafferty, *Anal. Chem.*, 1967, **39**, 1748.
[2] J. H. Beynon, R. G. Cooks, J. W. Amy, W. E. Baltinger and T. Y. Ridley, *Anal. Chem.*, 1973, **45**, 1023A.
[3] J. C. Porter, J. H. Beynon and T. Ast, *Org. Mass Spectrom.*, 1981, **16**, 101.
[4] R. G. Cooks, J. H. Beynon, R. M. Caprioli and G. R. Lester, *Metastable Ions*, Elsevier, Amsterdam, 1973.
[5] T. R. Kemp, in *Mass Spectrometry (Specialist Periodical Reports)*, Vol. 7, Royal Society of Chemistry, London, 1984, pp. 179–182.
[6] R. K. Boyd and J. H. Beynon, *Org. Mass Spectrom.*, 1977, **12**, 163.
[7] M. E. Rose, in *Mass Spectrometry (Specialist Periodical Reports)*, Vol. 7, Royal Society of Chemistry, London, 1984, pp. 259–264.
[8] J. L. Holmes and J. K. Terlouw, *Org. Mass Spectrom.*, 1980, **15**, 383.

# 11

# Ion–molecule reactions

## 11.1 DEFINITIONS

If the pressure in the ion source is greater than $10^{-5}$ mmHg, the mean free path of the ions in the ionization chamber decreases sufficiently for an appreciable fraction of them to give ion–molecule reactions by collision with the non-ionized molecules M:

$$X^+ + M \rightarrow Y^+$$

which may be followed by

$$Y^+ + M \rightarrow Z^+$$

where $X^+$, $Y^+$ and $Z^+$ are called primary, secondary and tertiary ions respectively.

## 11.2 EXPERIMENTAL METHODS

Ion–molecule reactions are used for the following purposes:

(a) to identify reaction paths;
(b) to determine equilibrium constants;
(c) to study reaction kinetics;
(d) to study the changes resulting from modifications in the excitation energy of the reactant ion;
(e) to study the internal energy of the reaction products;
(f) to determine reaction mechanisms.

Most studies have involved simple molecules that exist in the gaseous state at ambient temperature. Some of the types of apparatus used are described below.

### 11.2.1   High-pressure ion sources

The reactions take place in a modified ion source. The object of the modifications is to increase the pressure in the source, and hence increase the ionization efficiency (Fig. 11.1), to create equilibrium conditions, and to allow kinetic studies.

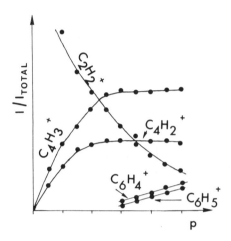

Fig. 11.1 — Effect of pressure in the ion source on the relative intensity of ions produced from acetylene. (Reproduced by permission, from P. S. Rudolph and C. E. Nalton, *J. Phys. Chem.*, 1959, **63**, 916. Copyright 1959, American Chemical Society, Washington D.C.)

The primary ions are formed by electron impact, photon bombardment, or use of a radioactive source. Increasing the pressure allows secondary and tertiary ions to be obtained by consecutive reactions, as shown above.

The main disadvantage of the method is that the ion species are formed in the source and therefore the relationship between the reactant ion and the product ion must be established by complementary studies.

### 11.2.2   Tandem spectrometers

The experimental methods are now sufficiently well developed to overcome the disadvantages of the high-pressure ion source. In the tandem methods, the region in which the reactant ion is produced is separated from the reaction region by a collision chamber (cf. Section 10.3.2). The primary ions formed in the ion source are first separated by a magnetic analyser, and the species of interest is introduced into the collision chamber, which is equipped with an inlet system for admission of a reactive gas. The secondary and/or tertiary ions formed in the collision chamber are then separated with a second magnetic analyser. The ions produced can be extracted from the collision chamber either collinearly with the primary ion beam (longitudinal method) or perpendicularly to it (transverse method, Fig. 11.2).

This method unambiguously establishes the parent–daughter or reactant–product relationship.

Some instruments have a very restricted reaction region, consisting of the zone of intersection of the primary ion beam with a molecular jet stream of reactive gas. Further details of ion–molecule reactions are given in Section 13.2.2.

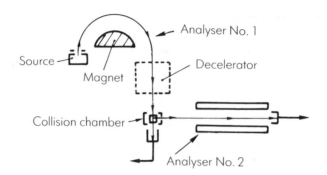

Fig. 11.2 — A transverse tandem spectrometer.

### 11.2.3 Measurement of angular distribution

Another new approach is employed to measure the angular and energy distributions of the reaction products (Fig. 11.3). Such experiments are crucial in establishing

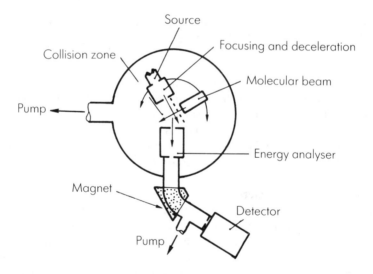

Fig. 11.3 — Apparatus for angular distribution measurements. (Reproduced, by permission, from Z. Herman, J. Kerstetter, T. Rose and R. Wolfgang, *Disc. Faraday Soc.*, 1967, **44**, 123. Copyright 1967, Royal Society of Chemistry, London.)

reaction mechanisms, and categorizing them as collisional complex-formation reactions or stripping reactions (Section 11.5).

## 11.3  TYPES OF ION–MOLECULE REACTION

The main types of ion–molecule reaction are outlined below, with typical examples. The primary ion is the reactant ion, and the secondary ion is the product ion.

### 11.3.1  Proton transfer

$$CH_4^+ + CH_4 \rightarrow CH_5^+ + CH_3$$
$$NH_3^+ + NH_3 \rightarrow NH_4^+ + NH_2$$
$$M + CH_5^+ \rightarrow MH^+ + CH_4$$

### 11.3.2  Charge transfer

$$He^+ + CH_4 \rightarrow He + CH_4^+ \text{ (non-dissociative)}$$
$$He^+ + CH_4 \rightarrow He + CH_3^+ + H^+ \text{ (dissociative)}$$

### 11.3.3  Association (collisional complex-formation)

$$
\begin{aligned}
Ar^+ + H_2 &\rightarrow ArH^+ + H \\
CH_3^+ + CH_4 &\rightarrow C_2H_5^+ + 2H \\
M + Na^+ &\rightarrow MNa^+ \\
NH_4^+ + NH_3 &\rightarrow (NH_4NH_3)^+ \\
M + NH_4^+ &\rightarrow MNH_4^+ \\
M + (NH_4NH_3)^+ &\rightarrow (MNH_4NH_3)^+ \\
M + C_3H_5^+ &\rightarrow (MC_3H_5)^+ \\
M + C_4H_9^+ &\rightarrow (MC_4H_9)^+ \\
O^+ + N_2 &\rightarrow NO^+ + N \\
H_2O^+ + H_2O &\rightarrow H_3O^+ + OH \\
H_3O^+ + H_2O &\rightarrow (H_3O)^+H_2O \\
(H_3O)^+H_2O + H_2O &\rightarrow (H_3O)^+(H_2O)_2 \\
\text{etc.} &\rightarrow (H_3O)^+(H_2O)_n
\end{aligned}
$$

### 11.3.4  Dissociation

$$CH_4^+ + Ar \rightarrow CH_3^+ + H + Ar, \ CH_2^+ + 2H + Ar$$

### 11.3.5  Ion–molecule reactions of heavy ions from solids

Many examples can be given of spectra containing ions of mass greater than that of the parent molecule. High molecular-weight organic compounds may yield mass spectra which exhibit peaks for ions formed from parent molecules by attachment of a proton, or a metal ion (cationization), or a fragment ion. For example, peptides may give rise to ions formed by a combination of the parent molecule and one of its fragment ions. In some cases these products are formed by ion–molecule reactions in the gas phase. Figure 11.4 depicts part of the mass spectrum for glucose thermally desorbed in the presence of a mixture of alkali-metal salts. The parent molecule is said to be 'cationized' (see Section 11.10.2).

The studies so far made have mainly aimed at establishing decay schemes for these ions. Collision experiments performed by introducing the ion species under

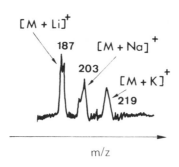

Fig. 11.4 — Mass spectrum of glucose. (Adapted, with permission, from R. Stoll and F. W. Röllgen, *Org. Mass Spectrom.*, 1981, **16**, 72. Copyright 1981, Heyden Ltd., London.)

study into a collision chamber are used for finding the structure of the ion. Thus for light ions, studies are made with gases and known ions to explore the reaction mechanism. For heavy ions, the studies concentrate on the practical and analytical aspects.

This difference in aims is due to the fact that the heating needed to vaporize heavy ions makes it difficult to establish equilibrium conditions in the ion source or the collision chamber, and it is also difficult to keep the pressure constant in the reaction region. However, an automatic direct introduction system has been designed which overcomes this problem (thermospray, Sections 4.7.2 and 13.2.3).

## 11.4   REACTION CROSS-SECTION

If the primary ions in the reaction chamber move along the $x$-axis and undergo reactive collisions with gas molecules, the decrease in the intensity ($I$) of the primary beam over a distance d$l$ is expressed by

$$\mathrm{d}I = \sigma n I_0 \mathrm{d}l$$

where $I_0$ is the initial intensity, $n$ the number of neutrals per unit volume and $\sigma$ the reaction cross-section. If the total distance over which reaction takes place is $l$, then for the primary ion the final beam intensity is

$$I_l = I_0\, e^{-\sigma n l}$$

If the value of $n\sigma l$ is $\ll 1$, then the intensity of the beam of secondary ions formed, $I_2$, can be approximated by $I_2 = I_0 n \sigma l$.

## 11.5   ENERGETIC AND KINEMATIC ASPECTS

These aspects can be dealt with by use of classical Newtonian theory. Consider a collision between an ion of mass $m_1$ with a molecule of mass $m_2$ to give an ion of mass

$m_3$ and a neutral species of mass $m_4$. If the initial velocities with reference to the laboratory are $\mathbf{v}_{1i}$ and $\mathbf{v}_{2i}$, the initial relative velocity of the reacting particles is $\mathbf{v}_{ri} = \mathbf{v}_{1i} - \mathbf{v}_{2i}$. The kinetic energy of these particles that will be available for conversion into the internal energy and kinetic energy of the products is the kinetic energy of the centre of mass:

$$T_{CMi} = M_r \mathbf{v}_{ri}^2 / 2$$

where $M_r$ is the reduced mass of the reactants at the point of collision, and is given by $M_r = m_1 m_2 / (m_1 + m_2)$. The velocity of the centre of mass, $\mathbf{v}_{CM}$, in the laboratory system, is given by $\mathbf{v}_{CM} = (m_1 \mathbf{v}_{1i} + m_2 \mathbf{v}_{2i}) / (m_1 + m_2)$. After the collision and separation of the final particles, the laboratory velocities of these particles will be $\mathbf{v}_{3f}$ and $\mathbf{v}_{4f}$, with final relative velocity $\mathbf{v}_{rf} = \mathbf{v}_{3f} - \mathbf{v}_{4f}$. These relationships are shown in the vector diagram in Fig. 11.5.

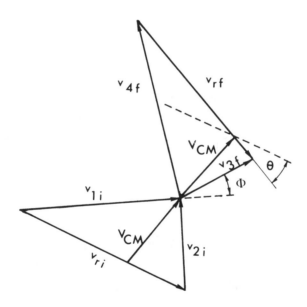

Fig. 11.5 — Vector diagram of particle velocities before and after collision. (Reproduced, by permission, from E. W. McDaniel, V. Čermák, A. Dalgarno, E. E. Ferguson and L. Friedman, *Ion–Molecule Reactions*, p. 118. Copyright 1970, John Wiley & Sons Inc., New York.)

Consider further the case if a complex of mass $m_1 + m_2$ is formed during the collision, before decomposing into the products. Its velocity will be $\mathbf{v}_{CM}$. The total internal energy of this complex, $E_{total}$, will be the sum of the heat of reaction, $W$, the initial internal excitation energy of the reactants, $E_i$, and the kinetic energy of the centre of mass:

$$E_{total} = T_{CMi} + W + E_i$$

If the velocity of the reactant ions is increased sufficiently, the value of $T_{CMi}$ can reach the dissociation energy $D$ and the complex will fragment. This will occur at a relative kinetic energy $T_{CMi, min}$ given by

$$T_{CMi, min} = D - (W + E_i)$$

where $D$ is equal to $E_{total}$. If $v_{2i}$ is negligible compared with $v_{1i}$, then

$$T_{CMi, min} = M_r v_{1i}^2/2 = D - (W + E_i)$$

and $T_{lab, min}$, the threshold kinetic energy for $m_1$ that is necessary for this fragmentation, is given by

$$T_{lab, min} = [D - (W + E_i)] (m_1 + m_2)/m_2$$

Formation of such a complex is an extreme case of the reaction mechanism, at very low energies. The recoil velocity $v_{3f}$ for the product ion can vary from zero (if all the energy available is used for excitation of the products) to a maximum when none of the energy is used for excitation. Since the ion produced (mass $m_3$) will have effectively the final centre of mass velocity ($v_{4f}$ will be negligible in comparison), its final velocity $v_{3f}$ can be calculated from

$$v_{3f} = \sqrt{2T_{CMf}m_4/(m_3m_4 + m_3^2)}$$
$$= \sqrt{2(T_{CMi} + W + E_i - E_f)m_4/(m_3m_4 + m_3^2)}$$

where $E_f$ is the final internal excitation energy of the products [1].

### 11.5.1   The Newton diagram
The final relative velocity vector may take any direction in space, but its magnitude is restricted by the conservation of energy. This can be shown schematically by construction of a 'Newton diagram' as in Fig. 11.6. For each value of the final relative translational energy, the recoil vectors for product $m_3$ can range over a sphere of radius $v_{rf}m_4/M_f$, where $M_f$ is the reduced mass of the products. The angle $\theta$ between $v_{ri}$ and $v_{rf}$ describes the angular distribution, which has cylindrical symmetry about $v_{ri}$. Figure 11.6 shows the relationships for two values of the final relative translational energy, circle 1 corresponding to the maximum value.

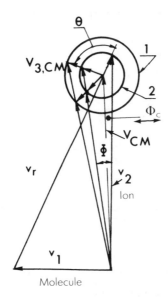

Fig. 11.6 — A Newton diagram. (Reproduced, by permission, from E. W. McDaniel, V. Čermák, A. Dalgarno, E. E. Ferguson and L. Friedman, *Ion–Molecule Reactions*, p. 120. Copyright 1970, John Wiley & Sons Inc., New York.)

### 11.5.2 Complex-formation and stripping

In principle, an ion–molecule reaction can take place in two ways:

(a) by formation of a transitory complex;
(b) by a so-called stripping mechanism.

In the first, the complex formed has a lifetime that is no longer than the period of rotation of the complex. When the complex decomposes, the angular distribution of the reaction products is isotropic in the centre of mass system, i.e. it is symmetrical around the direction of $v_{ri}$. Because the orbital angular momentum vector must be perpendicular to the plane of the velocity vectors of the reactants, the density of the recoil vectors of the products will be maximal at the poles of the sphere of distribution of the vectors (Section 11.5.1) resulting in two peaks on the angular distribution plot (Fig. 11.7, which corresponds to the Newton diagram in Fig. 11.6). This is the so-called 'glory' effect. The interaction takes place between the whole molecule M and reactive ion $M_x^+$.

In the second mechanism, in contrast, the interaction takes place between the whole molecule and only part of the ion (or vice versa). The remainder of the partially reactive species behaves as a 'spectator', and this mechanism is called spectator stripping. No momentum is exchanged between the spectators and the reacting particles, and the spectator group retains the velocity vector of its parent. The product of the reaction will have a momentum that is the sum of its original momentum and that of the reactive entity split off from the spectator group, and as

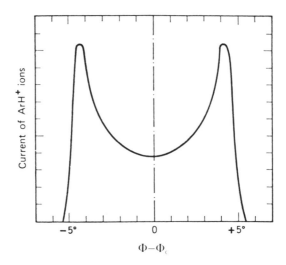

Fig. 11.7 — Angular distribution of $ArH^+$ ions for an isotropic distribution of $v_{3,CM}$ recoil vectors. (Reproduced, by permission, from E. W. McDaniel, V. Čermák, A. Dalgarno, E. E. Ferguson and L. Friedman, *Ion–Molecule Reactions*, p. 120. Copyright 1970, John Wiley & Sons Inc., New York.)

shown in the angular distribution diagram in Fig. 11.8, in the stripping mechanism there is an anisotropic angular distribution of the product ion.

In general, the stripping mechanism would be favoured by reactions between species with high relative velocities, and the complex-formation mechanism at low relative velocities. It should be noted, however, that both mechanisms have been challenged as oversimplified, and new mechanisms advanced [2].

### 11.6 PROTON TRANSFER [3]

Let us consider the proton transfer reaction

$$B_1H^+ + B_2 \rightarrow B_1 + B_2H^+$$

where $B_1$ and $B_2$ are atoms or groups of atoms. If equilibrium is achieved in the reaction chamber, then

$$-RT \ln K = \Delta G^\circ \tag{11.1}$$

where $K$ is the equilibrium constant

$$K = [B_1][B_2H^+]/[B_1H^+][B_2]$$

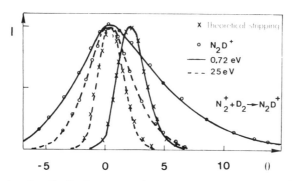

Fig. 11.8 — Angular distribution of $N_2D^+$ formed by reaction $N_2^+ + D^2 \rightarrow N_2D^+ + D$. (Reproduced, with permission, from Z. Herman, J. Kerstetter, T. Rose and R. Wolfgang, *Disc. Faraday Soc.*, 1967, **44**, 123. Copyright 1967, Royal Society of Chemistry, London.)

and $R$, $T$ and $\Delta G^\circ$ have their usual thermodynamic meaning. $[B_1]$ and $[B_2]$ can be expressed as the partial pressures, and $[B_1H^+]$ and $[B_2H^+]$ as the ion currents. Further, $\Delta G^\circ$ can be expressed in terms of $\Delta H^\circ$ and $\Delta S^\circ$

$$\Delta G^\circ = \Delta H^\circ - T\Delta S^\circ \tag{11.2}$$

the latter being approximated by the change in rotational entropy ($\sigma$):

$$\Delta S^\circ \sim \Delta S^\circ_{\text{rot}} = R \ln [\sigma(B_1H^+)\sigma(B_2)/\sigma(B_1)\sigma(B_2H^+)] \tag{11.3}$$

Equation (11.1) gives $\Delta G^\circ$, which in combination with Eqs. (11.2) and (11.3) gives $\Delta H^\circ$, from which the proton affinity $A_p(B_2)$ of $B_2$ can be calculated if that of $B_1$ is known:

$$\Delta H^\circ = A_p(B_1) - A_p(B_2) \tag{11.4}$$

The experimental results allow calculation of a scale of proton affinities in the gas phase from the proton affinities of $H_2$, $H_2O$ and $NH_3$ as reference values.

Thus $CH_4$ is found to have a proton affinity between those of $H_2$ and $H_2O$, namely 128.2 kcal/mole. Other molecules with proton affinities (kcal/mole) in this region include $CH_3CHO$ (185), $C_2H_5COOH$ (189.5) and $H_2S$ (173.9).

## 11.7 REACTIONS WITH METAL IONS

### 11.7.1 Gas-phase reactions catalysed by $Fe^{3+}$ [4]
Iron(III) ions can catalyse certain gas phase reactions, for example:

$$Fe^{3+} + N_2O \rightarrow FeO^+ + N_2$$

$$\underline{FeO^+ + CO \rightarrow Fe^{3+} + CO_2}$$

$$N_2 + CO \rightarrow N_2 + CO_2 \quad \Delta H^\circ = -107 \text{ kcal/mole}$$

### 11.7.2　Reactions of $Cu^+$ ions [5]

A typical scheme for a copper(I)-catalysed reaction is that for the formation of ethylene from ethyl chloride

$Cu^+$ +　(a)

$Cu^+$ ... $Cl^-$ + ... H　(b)

$Cl$–$H$ / $Cu^+$　(c)

CuCl +

Transfer of Cl

$HCl$ .... $Cu^+$ +

Elimination of
alkene

$HCl$ + $Cu^+$

Elimination of HCl

The same types of reaction have been proposed for the catalytic effects of $Al^+$ [6], $Mg^+$ and $Li^+$ [7] on aliphatic alcohols and alkyl halides. In general, the reactions between an $M^+$ ion and an HXY molecule involve the following steps:

(a) association of the ion with the molecule;
(b) rearrangement;
(c) transfer of X;
(d) elimination of HX;
(e) dissociation and formation of $MOH^+$ or $MY^+$.

The system involving $Al^+$ and $C_2H_5Cl$ can be represented by the diagram shown in Fig. 11.9.

### 11.7.3　Reactions of $Rh^+$ ions [8]

$Rh^+$ ions catalyse the dehydrogenation of alkanes, the probable reaction paths being those shown opposite.

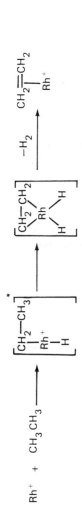

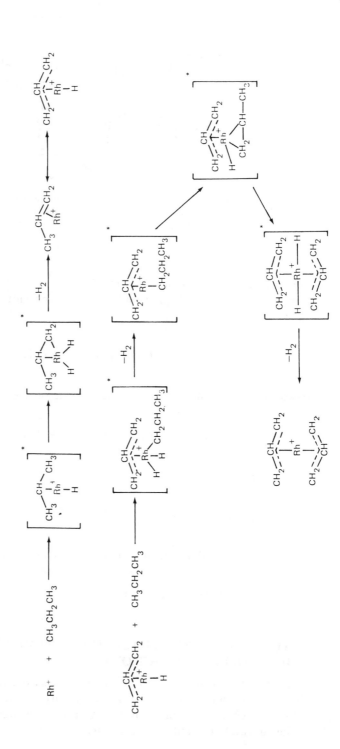

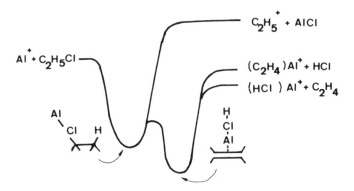

Fig. 11.9 — Energy-level diagram for the aluminium–ethyl chloride system. (Reproduced by permission, from R. V. Hodges, P. B. Armentrout and J. L. Beauchamp, *Intern. J. Spectrom. Ion Phys.*, 1979, **29**, 375. Copyright 1979, Elsevier, Amsterdam.)

### 11.7.4   Ligand exchange

An alternative reaction path is possible in the copper(I)-catalysed formation of ethylene from ethyl chloride (Section 11.7.2), starting with the reaction intermediate (c) shown in Section 11.7.2:

$$Cu(EtCl)^+ + EtCl \rightarrow (CuEtCl)(HCl)^+ + C_2H_4$$

followed by ligand exchange with another EtCl molecule

$$(CuEtCl)(HCl)^+ + EtCl \rightarrow Cu(EtCl)_2^+ + HCl$$

which may be formalized as

$$(CuLL')^+ + L \rightarrow (CuL_2)^+ + L'$$

For $Cu^+$ ions, the binding energies for various ligands increase in the order

$$MeCN > C_2H_4 > EtCl > HCl$$

### 11.8   STUDY OF THE EXCITED STATES OF PRODUCTS FORMED BY ION–MOLECULE REACTIONS

In exothermic ion–molecule reactions the products (ions or neutral species) can be produced in various excited rotational, vibrational or electronic states. Experimental detection of these states requires the use of sufficiently intense ion beams, which can be achieved with lasers and molecular jets.

### 11.8.1  Infrared chemiluminescence [9,10]

*Vibrational levels of HCl*

Figure 11.10 shows the apparatus used for detecting the vibrational levels of HCl, produced by the reaction of chloride ions with hydrogen bromide or hydrogen iodide molecules, the chloride ions being produced from carbon tetrachloride by electron impact.

$$Cl^- + HBr \rightarrow HCl \ (v = 0, 1, \ldots) + Br^-$$

$$Cl^- + HI \ \ \rightarrow HCl \ (v = 0, 1, \ldots) + I^-$$

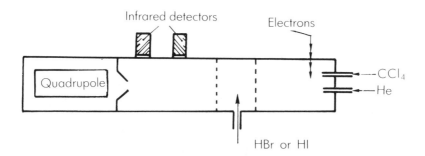

Fig. 11.10 — Principle of determination of vibrational energy levels of HCl. (Adapted, by permission, from T. S. Zwier, V. M. Bierbaum, G. B. Ellison and S. R. Leone, *J. Chem. Phys.*, 1980, **72**, 5426. Copyright 1980, American Physical Society, New York.)

*Study of dimers*

Another example is the study of excited states of dimers: $(C_2H_2)_2$, $(C_2H_4)_2$, and $(N_2O)_2$ have been studied in this way. Their infrared spectra in the region 3000–3300 cm$^{-1}$, the domain for the C–H vibrations, were obtained. For acetylene and methylacetylene, rotational fine structure was observed in the vibrational spectrum.

The apparatus used (shown schematically in Fig. 11.11) consisted of a molecular

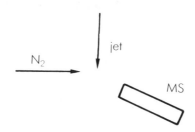

Fig. 11.11 — Principle of study of dimers

jet of the sample and a laser beam to excite the spectrum of the neutral aggregates, and establish their nature. A mass spectrometer was used to determine their mass.

Another interesting study was that of predissociation of the $(C_2H_2)_2$ and $(N_2O)_x$

aggregates ($x = 2$–100). The mass analysis was done with a time-of-flight spectrometer or a quadrupole analyser. For the $N_2O$ aggregates, the spectral region of the $v_1 + v_2$ band of the monomer was examined. The results obtained at different pressures showed a shift of the bands with increase in the size of the aggregates. This displacement could be correlated with the transition from the gaseous state to the solid state. The solid state characteristics first appeared at a degree of polymerization between $(N_2O)_{20}$ and $(N_2O)_{60}$, and calculations based on a semi-classical approximation located the transition at $x = 55$. The duration of the predissociation state was about $10^{-4}$ sec.

### 11.8.2  Ultraviolet and visible chemiluminescence [10]

With an apparatus of the type shown in Fig. 11.12, $CO^+$ was shown to give

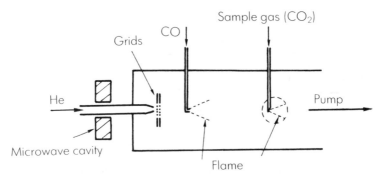

Fig. 11.12 — Detection of the chemiluminescence of $CO^+$ ions. (Adapted, by permission, from M. Tsuji, T. Susuki, M. Mizukami and Y. Nishimurs, *J. Chem. Phys.*, 1985, **83**, 1677. Copyright 1985, American Physical Society, New York.)

chemiluminescence in the region 200–800 nm. The luminescence was produced by exciting helium in a microwave cavity at a frequency of 2.45 GHz, resulting in $He(2^3S)$, $He^+$ and $He_2$ (mainly $He^+$), and admitting it into the reaction chamber, into which carbon monoxide was introduced. The reaction of $He^+$ with CO resulted in $C^+$ ions, which then underwent reaction with $CO_2$ admitted through the sample gas inlet, to produce $CO^+$. The chemiluminescence of the $CO^+$ was detected with a photomultiplier and photon counter. Figure 11.13 shows the spectrum corresponding to the transition

$$^2P_u(C^+) + CO_2(X\Sigma_g^+) \rightarrow CO(X^2\Sigma^+)$$

It has been found that it is chemiluminescence from $CO^+$ ions that makes comet tails visible.

### 11.9  REACTIVITY OF IONS AS A FUNCTION OF THEIR INTERNAL ENERGY

The reactivity of ions depends on their energy at the time of reaction. As examples will consider the reactions of $HOC^+$ and $O(^3P)$ with other species.

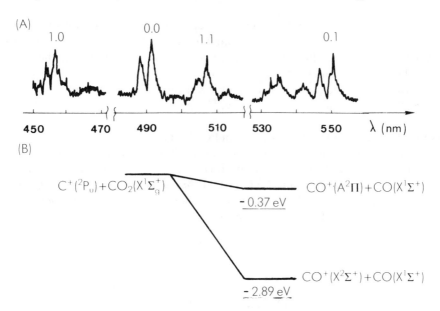

Fig. 11.13 — (A) Spectrum of chemiluminescence of $CO^+$ ions; $A \rightarrow X$ transition; (B) correlation of the $C_s$ symmetry between the reactants and products. (Adapted, by permission, from M. Tsuji, T. Susuki, M. Mizukami and Y. Nishimurs, *J. Chem. Phys.*, 1985, **83**, 1677. Copyright 1985, American Physical Society, New York.)

### 11.9.1   $HOC^+ + H_2$

$HOC^+$ ions play an important role in the chemistry of flames, plasmas and interstellar gas. They exist in two isomeric forms, $HOC^+$ and $HCO^+$, and appear after reaction between $H_3^+$ and CO:

$$H_3^+ + CO \begin{cases} \nearrow 6\% \ HOC^+ \\ \searrow 94\% \ HCO^+ \end{cases}$$

A study of the concentrations of these two ions in 14 sources of interstellar gas has shown, however, the concentration ratio is $HCO^+/HOC^+ = 330$, and this apparent anomaly is explained by the existence of the catalytic conversion of $HOC^+$ into $HCO^+$ by the reactions

$$HOC^+ + H_2 \rightarrow HCO^+ + H_2$$
$$HOC^+ + H \ \rightarrow HCO^+ + H$$

Proof of this isomerization has been obtained by forming $HOC^+$ ions from $CD_3OH$ by electron impact, extracting these ions (which have an internal energy between 0 and 1.2 eV) from the ionization chamber, separating them by cyclotron resonance and injecting them into a reaction chamber (a second cyclotron resonance cell) where they undergo collisional reaction with hydrogen. Analysis of the products shows the formation of HCO$^+$, It may be remarked that the pressure of the reagent gas (hydrogen) in the cyclotron resonance cell can be kept low, since the reactant ions can be kept in the cell for a relatively long time.

### 11.9.2   Reactions of O($^3P$) with C$_6$H$_6$ and C$_6$D$_6$ [11]

The triplet state of oxygen, O($^3P$), can be formed by excitation with a radiofrequency generator. The products of its reaction with benzene have been analysed with a time-of-flight analyser and shown to be C$_6$H$_5$OH (*m/z* 94), C$_6$H$_5$O (*m/z* 93), C$_5$H$_6$ (*m/z* 66), C$_5$H$_5$ (*m/z* 65), together with various other species; the energy levels of the various components of the system are shown in Fig. 11.14. The reactions of O($^3P$)

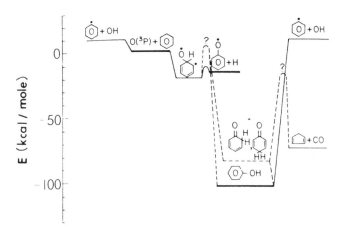

Fig. 11.14 — Energy levels for the O($^3P$) plus benzene system. (Adapted, by permission, from S. J. Sibener, R. J. Buss, P. Casavecchia, T. Hirooka and Y. T. Lee, *J. Chem. Phys.*, 1980, **72**, 4341. Copyright American Physical Society, New York.)

with benzene are important in the study of reactions in the atmosphere, particularly those involved in photochemical pollution and combustion products.

## 11.10   ANALYTICAL APPLICATIONS OF ION–MOLECULE REACTIONS

### 11.10.1   Chemical ionization

Proton transfer and addition reactions are used for identification of molecules. The molecule is ionized by means of primary ions from a reagent gas, which can transfer a proton to the molecule or form adduct ions with it. This technique is called chemical ionization and is discussed in detail in Section 13.2. It can be regarded as a gas-phase acid–base reaction.

### 11.10.2   Cationization

The mass spectra obtained for solid substances by laser desorption and ionization, ion or atom bombardment, or field desorption and ionization may contain peaks for the species produced by attachment of a metal ion to the parent molecule. This type of quasi-molecular ion is very important in the determination of molecular weight. An example is shown in Figs. 11.15 and 11.16, for the cationization of glucose. The metal ions can arise from traces in the original sample, or from compounds deliberately added to it.

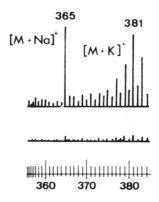

Fig. 11.15 — Part of mass spectrum of sucrose, showing peaks for cationized species. (Reproduced by permission, from E. Constantin, *Org. Mass Spectrom.*, 1982, **17**, 346. Copyright 1982, Heyden Ltd, London.)

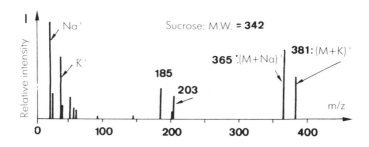

Fig. 11.16 — Mass spectrum of sucrose, showing cationization.

### 11.10.3   Charge transfer

Charge transfer is used to obtain neutral particles with a clearly defined energy. A beam of neutrals (M) of this type is created by (a) forming $M^+$ ions, (b) accelerating them to the required energy, then (c) transferring their charge to molecules of M. The beam of neutrals in the gun of a fast atom bombardment (FAB) source is obtained in this way (see Section 13.3).

### 11.10.4 Collision chamber reactions

The study of decomposition of metastable ions is very important in establishing molecular structures. To increase the intensities of the metastable peaks, the metastable ions are introduced into a collision chamber in a field-free region, where their fragmentation is induced by collisions with a 'collision gas', as in the MIKES, $B/E$ and $B^2/E$ techniques (Sections 10.3.2, 10.3.3, 17.5).

**REFERENCES**

[1] E. W. McDaniel, V. Čermák, A. Dalgarno, E. E. Ferguson and L. Friedman, *Ion–Molecule Reactions*, Wiley–Interscience, New York, 1970, p. 119.
[2] S. A. Safron, in *Mass Spectrometry* (*Specialist Periodical Reports*), Vol. 7, Royal Society of Chemistry, London, 1984, pp. 82–94.
[3] P. Kebarle, *Ann. Rev. Phys. Chem.*, 1977, **28**, 445.
[4] M. M. Kappes and R. H. Staley, *J. Am. Chem. Soc.*, 1981, **103**, 1286.
[5] R. W. Jones and R. H. Staley, *J. Am. Chem. Soc.*, 1980, **102**, 3794.
[6] R. V. Hodges, P. B. Armentrout and J. L. Beauchamp, *Intern. J. Spectrom. Ion Phys.*, 1979, **29**, 375.
[7] R. D. Wieling, R. H. Staley and J. L. Beauchamp, *J. Am. Chem. Soc.*, 1975, **97**, 924.
[8] G. D. Byrd and B. S. Freiser, *J. Am. Chem. Soc.*, 1982, **104**, 5944.
[9] T. S. Zwier, V. M. Bierbaum, G. B. Ellison and S. R. Leone, *J. Chem. Phys.*, 1980, **72**, 5426.
[10] M. Tsuji, T. Susuki, M. Mizukami and Y. Nishimura, *J. Chem. Phys.*, 1985, **83**, 1677.
[11] S. J. Sibener, R. J. Buss, P. Casavecchia, T. Hirooka and Y. T. Lee, *J. Chem. Phys.*, 1980, **72**, 434.

# 12

# Ion clusters

## 12.1  INTRODUCTION

An ion cluster is a large ion-molecule of variable size, formed from an ion and one or more molecules. The experimental study of ion clusters in the gas phase is linked to developments in mass spectrometry. The existence of ion clusters was detected in high-pressure ion sources during studies on ion–molecule reactions (see Section 11.3.3). Clusters also arise when SIMS, ion-bombardment, laser irradiation, field desorption and fission-fragment bombardment (Chapter 13) are used. $H^+(H_2O)_n$ ion clusters have been detected in the ionosphere and in the stratosphere.

Mass spectrometry is the only technique currently available for characterizing clusters in terms of composition and evolution, and for providing thermodynamic and kinetic data.

The study of ion clusters is important not only for their characterization but also, by extrapolation, for study of the liquid and solid states. The following aspects need to be considered in this respect: the mechanism of formation and the lifetime of a cluster, deactivation by collision or exothermic chain reactions, the distribution of the size of the cluster and its redistribution after loss of a component molecule, and the radius of the cluster as a function of the number of molecules surrounding the central ion.

## 12.2  EXPERIMENTAL METHODS FOR STUDY OF CLUSTERS

The specific conditions necessary for the study of clusters are obtained mainly in four types of equipment:

(a)  classical high-pressure mass spectrometers;

(b) flowing afterglow or drift-tube units;
(c) cyclotron resonance spectrometers;
(d) molecular jets.

### 12.2.1   High-pressure ion-source spectrometers
In these instruments the ion source functions under equilibrium conditions, in which the primary ions (the nucleation centres) are produced by various means. Metal ions, for example, are produced by thermal ionization. The mass analysis is done with a magnetic or quadrupole analyser.

### 12.2.2   Flowing afterglow and drift-tube methods [1–3]
In the flowing afterflow method, the primary ions are produced by electron impact or discharge ionization, then transported by a carrier gas through an electric field, where they move in 'swarms' at rates determined by their mass and charge and the laws of diffusion. They may then be allowed to react with a suitable reactant gas, and the products are identified by means of a mass analyser (e.g. a quadrupole analyser as in Fig. 12.1).

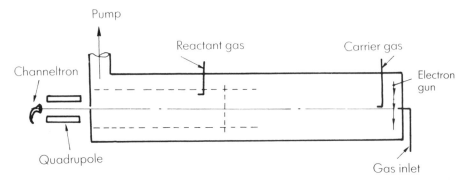

Fig. 12.1 — Scheme of a flowing afterflow apparatus.

In the 'SIFT' (selected ion flow tube) method, the parent ions of the aggregates are formed by electron impact or charge transfer ionization, separated by a mass analyser (e.g. a quadrupole analyser, Fig. 12.2) and introduced into a reaction chamber through which a reactive gas is passed, and the product ions are extracted by a weak electric field and analysed by a second quadrupole analyser.

### 12.2.3   Cyclotron resonance method
In this technique the ion cyclotron resonance analyser (Section 5.3) is used to trap the parent ions (formed by electron impact or soft ionization techniques) in the presence of a reactant gas. The main advantage of the technique is that the reactant ion can be held for relatively long periods in the reaction zone, thus enhancing the probability of reaction. The method is particularly useful in the study of low-energy collisions.

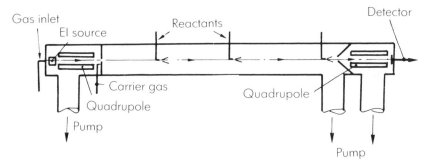

Fig. 12.2 — Scheme of the SIFT apparatus. (Reproduced, by permission, from M. Tichy, A. B. Rakshit, D. G. Lister, N. D. Twiddy, N. G. Adams and D. Smith, *Intern. J. Mass Spectrom. Ion Phys.*, 1979, **29**, 231. Copyright 1979, Elsevier, Amsterdam.)

### 12.2.4 The supersonic jet method

The principle of supersonic jets was first suggested in 1951, but its practical realization was rather slow [4]. In this method the sample stream is expanded into a low-pressure region through a small orifice, resulting in a supersonic jet with high Mach number. A 'skimmer' (a cone with a small orifice) is used to withdraw the sample from the supersonic stream into the ionization region, the remainder being pumped away. Figure 12.3 shows a typical arrangement.

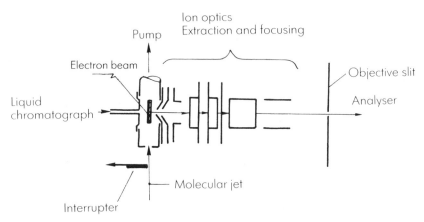

Fig. 12.3 — A supersonic jet apparatus. (Adapted, by permission, from K. Stephan, J. H. Futrell, K. I. Peterson, A. W. Castleman, Jr., H. E. Wagner, N. Djuric and T. D. Märk, *Intern. J. Mass Spectrom. Ion Phys.*, 1982, **44**, 167. Copyright 1982, Elsevier, Amsterdam.)

## 12.3   DETERMINATION OF HEAT OF REACTION

Clusters are formed by a series of addition reactions between the ion ($I^+$) and the molecules (M):

$$I^+ + M \rightarrow IM^+$$

$$IM^+ + M \rightarrow IM_2^+$$

$$(IM_{n-1})^+ + M \rightarrow (IM_n)^+$$

If equilibrium is reached in the reaction region, then

$$(IM_{n-1})^+ + M \rightleftharpoons (IM_n)^+$$

with an equilibrium constant $K$

$$K = [(IM_n)^+]/[(IM_{n-1})^+][M]$$

where $[(IM_{n-1})^+]$ and $[(IM_n)^+]$ can be expressed in terms of their ion currents. Then (cf. Section 11.6), from

$$\Delta G^\circ = -RT \ln K$$

$$\Delta G^\circ = \Delta H^\circ - T\Delta S^\circ$$

$$\Delta S_{rot} = R \ln [\sigma(IM_{n-1})^+ \sigma(M)/\sigma(IM_n)^+]$$

the heat of reaction (a) and the heat of formation of a given species (b) can be calculated by use of the relationships

(a) $\Delta H^\circ = \Delta G^\circ + T\Delta S^\circ$
(b) $\Delta H^\circ = \Delta H^\circ(IM_n)^+ - \Delta H^\circ(M) - \Delta H^\circ(IM_{n-1})^+$

Two systems may be considered, as examples.

$A$. In the system $NH_3 + H_2O$, clusters of the type $NH_4(H_2O)_n^+$ are formed:

$$NH_4^+ + H_2O \rightarrow NH_4(H_2O)^+$$

$$NH_4(H_2O)^+ + H_2O \rightarrow NH_4(H_2O)_2^+$$

$$NH_4(H_2O)_{n-1}^+ + H_2O \rightarrow NH_4(H_2O)_n^+$$

The heats of formation of successive aggregates are reported to be as follows [5]:

| $(n, n-1)$ | (1,0) | (2,1) | (3,2) | (4,3) | (5,4) |
|---|---|---|---|---|---|
| $\Delta H^\circ_{(n,n-1)}$, kcal/mole | 17.2 | 14.7 | 13.2 | 12.2 | 9.7 |

$B$. In the system $Sr^+ + H_2O$, clusters are formed by reactions of the type

$$Sr(H_2O)_{n-1}^+ + H_2O \rightarrow Sr(H_2O)_n^+$$

with typical heats of formation as follows [6]

| $(n, n-1)$ | (1,0) | (2,1) | (5,4) |
|---|---|---|---|
| $\Delta H^\circ_{(n,n-1)}$, kcal/mole | 34.5 | 30.5 | 20.6 |

The results can conveniently be displayed graphically, as in Fig. 12.4. As might be expected, $\Delta H^\circ_{(n,n-1)}$ decreases with increasing $n$, i.e. with increasing size of the cluster.

## 12.4 STUDY OF THE STRUCTURE OF CLUSTERS

### 12.4.1 Clusters of the type $H^+(H_2O)_n$ [7,8]

The distribution of the mass spectrum intensities for $H^+(H_2O)_n$ ions leads to the conclusion that there is a high probability that clusters corresponding to $n = 21$ will

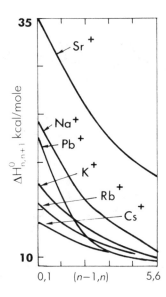

Fig. 12.4—Enthalpy differences between various cluster ions. (Adapted by permission, from I. N. Tang, M. S. Lian and A. W. Castleman, Jr., *J. Chem. Phys.*, 1976, **65**, 4022. Copyright, 1976, American Physical Society, New York.)

form. Studies of the clusters show that the abundance of the $n = 22$ cluster is considerably less than that of the $n = 21$ species, and a similar difference occurs in the case of the $n = 28$ and $n = 29$ clusters. Smaller effects are observed at $n = 26$ and $n = 30$. These observations are compatible with the hypothesis of an $H_3O(H_2O)_3^+$ ion with the structure shown in Fig. 12.5 and the formation of clusters with the dodecahedral structure shown in Fig. 12.6. It is proposed that proton transfer takes place within the cage structure at a rate greater than that of the molecular rotation. Thus the $H_3O^+$ formed at any instant of time would be in direct interaction with the water molecules surrounding it and in indirect interaction with the others. The stability of the cluster with $n = 21$ would then be attributable to the $H^+(H_2O)_{20}$ dodecahedron forming a clathrate cage around a neutral water molecule, the charge on the proton being distributed over the entire cage.

However, the structure of these gas phase clusters is far from being definitively established. Besides the experimental work, theoretical studies have been made, by quantum mechanics or Monto Carlo methods, or both. Recent calculations [8] have suggested the existence of a primary sheath of three or four molecules of water around a central $H_3O^+$ ion (Fig. 12.7) where the fundamental interaction would be exchange of two water molecules, one moving towards the $H_3O^+$ ion and the other away from it. In the gas phase this would lead to a structure different from that of an ice crystal. The calculations indicate that a third water sheath would begin to form before a second sheath was completed (Fig. 12.8).

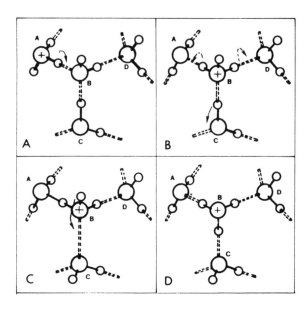

Fig. 12.5 — Structure of $H_3O^+(H_3O)_3$ clusters. (Adapted, by permission, from P. M. Holland and A. W. Castleman, Jr., *J. Chem. Phys.*, 1980, **72**, 5984. Copyright 1980, American Physical Society, New York.)

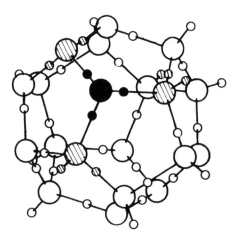

Fig. 12.6 — Structure of an $H_3O^+(H_2O)_n$ clathrate, showing a possible position for an $H_3O^+(H_2O)_3$ group (shaded). (Adapted, by permission, from P. M. Holland and A. W. Castleman, Jr., *J. Chem. Phys.*, 1980, **72**, 5984. Copyright 1980, American Physical Society, New York.)

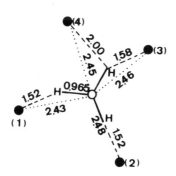

Fig. 12.7 — Structure of an $H_3O^+(H_2O)$ group. (Reproduced by permission, from E. Kochanski, *J. Am. Chem. Soc.*, 1985, **107**, 7869. Copyright 1985, American Physical Society, New York.)

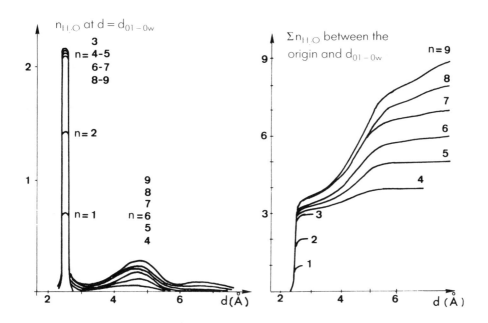

Fig. 12.8 — Numbers of water molecules in hydration sheaths of $H_3O^+$; $d_{O1-Ow}$ is the distance between the oxygen atom of $H_3O^+$ and the oxygen atom of an $H_2O$ molecule. (Reproduced by permission, from E. Kochanski, *J. Am. Chem. Soc.*, 1985, **107**, 7869. Copyright 1985, American Physical Society, New York.)

### 12.4.2 Ammonia clusters [9]

These clusters are important representatives of hydrogen-bonded systems. Also, because ammonia is present in the troposphere and in the atmosphere of certain planetary satellites, it is important to know the evolution, behaviour and structure of these clusters.

Molecular jet experiments show the formation of both protonated and non-protonated ion clusters of ammonia, $(NH_3)_nH^+$ and $(NH_3)_n^+$. Further, it has been observed that when the ratios of the ion currents of these species are plotted semilogarithmically as a function of $n$, maxima occur at certain 'magic' values of $n$. In a plot of log $[(NH_3)_n^+/(NH_3)_nH^+]$ against $n$ there is a maximum at $n = 4$ and minima at $n = 6$ and 21.

This behaviour can be explained as due to proton transfer within an $(NH_3)_n^+$ cluster, which may be followed by a dissociation step:

$$(NH_3)_n + h\nu \text{ (or } e^-) \rightarrow (NH_3)_n^+ + e^- \text{ (or } 2e^-)$$
$$\text{(proton transfer)}$$
$$(NH_3)_{n-2}NH_4^+ \ldots NH_2 \rightleftharpoons (NH_3)_{n-2}NH_4^+ + NH_2$$
$$\text{(dissociated)}$$

The ratio of $(NH_3)_n^+$ to $(NH_3)_{n-2}NH_4^+$ will depend on the degree of dissociation of the proton-transfer complex.

According to this hypothesis, the structure of the ion $(NH_3)_5^+$ is probably

$$
\begin{array}{c}
NH_3 \\
\vdots \\
H \\
| \\
H_3N \ldots H - N^+ - H \ldots NH_2 \\
| \\
H \\
\vdots \\
NH_3
\end{array}
$$

By analogy with $(NH_3)_4NH_4^+$ this ion should be very stable. The internal transfer of a proton will generate the ion $(NH_3)_3NH_4^+$ as follows:

$$(NH_3)_3NH_4^+ \ldots NH_2 \rightarrow (NH_3)_3NH_4^+ + NH_2$$

If a further molecule of ammonia is added (making $n = 6$), it will occupy an empty outer orbital to form

$$[(NH_3)_3NH_4^+ NH_2]NH_3$$

and the system will be less stable.

It may be observed that the cluster $(NH_3)_{20}NH_4^+$ is analogous to $(H_2O)_{20}H_3O^+$. A possible energy-level diagram is given in Fig. 12.9.

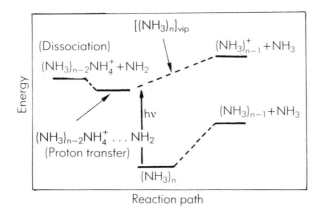

Reaction path

Fig. 12.9 — Energy levels for the $NH_3$ aggregate system (vip = vertically ionized product). (Reproduced by permission, from H. Shinahara, N. Nishi and N. Washida, *J. Chem. Phys.*, 1985, **83**, 1939. Copyright 1985, American Physical Society, New York.)

### 12.4.3   Carbon clusters [10–12]

Studies have been made of carbon clusters emitted from solid graphite under laser irradiation in a pulsed current of helium. The mixture of helium and graphite vapour is expanded into a vacuum in a molecular jet system, the supersonic nozzle beam being ionized by a laser beam. The products are then mass-analysed with a time-of-flight spectrometer.

An interesting result of these studies was the discovery of a very stable ionic species consisting of a cluster of 60 carbon atoms [10]. The experiments showed the influence of thermal equilibrium on the intensity of the peak for $C_{60}$. When the laser acts on the graphite target after several helium pulses, the mass spectrum contains peaks for clusters with various even numbers of carbon atoms in them, but the peak for $C_{60}$ is the dominant one. Under these experimental conditions collisional de-excitation will be minimal. If the laser pulse is synchronized with the helium

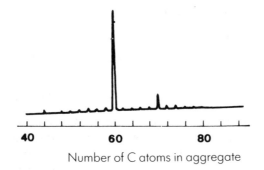

Number of C atoms in aggregate

Fig. 12.10 — Mass spectral peak for $C_{60}$. (Reproduced by permission, from *Nature*, Vol. 318, pp. 162–163. Copyright © 1985 Macmillan Magazines Ltd.)

intensity of the helium pulse, however, the peak for $C_{60}$ becomes much larger than the others (Fig. 12.10) as it also does if the interval of time between the laser pulse and the expansion is increased (since in both cases there is more time for thermal equilibrium to be reached).

The structure proposed for the $C_{60}$ cluster corresponds to a polygonal surface with 60 vertices and 32 faces, 12 of which are pentagonal and the other 20 are hexagonal (Fig. 12.11). The unit of surface structure is thus a pentagon surrounded by five hexagons.

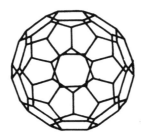

Fig. 12.11 — Proposed structure for $C_{60}$.

This supermolecule ('footballene') is spherical, aromatic and in the form of a truncated icosohedron. It is 7 Å in diameter and has an inner cavity 5 Å in diameter. This type of structure has filled molecular orbitals of a particularly stable nature.

These carbon clusters form stable compounds with lanthanum atoms [11], the predominant one being $C_{60}La$, but other stable aggregates are also formed, such as $C_{44}La$ and $C_{76}La$. They probably have the lanthanum atom inside the central cavity of the carbon structure. A polemic on this topic has developed [12].

The $C_{60}$ clusters do not appear to react with oxygen or water vapour.

## 12.5 REACTIVITY OF CLUSTERS [13]

### 12.5.1 Proton transfer [14]

Consider the case of the protonated water cluster $H_3O^+(H_2O)_n$ and its reaction with a base B:

$$H_3O^+(H_2O)_n + B \rightleftharpoons BH^+(H_2O)_m + (n - m + 1)H_2O$$

The value of $m$ will depend on the ability of the resulting ion $BH^+(H_2O)_m$ to retain water molecules. The heat of the reaction will depend on the difference between the proton affinities of the base and of water.

Attack by B can take place on different parts of the cluster, B accepting the proton and one or more molecules of water. It is very likely that at the start of the reactions B will not be in the correct position to become the centre of the $BH^+(H_2O)_n$ ion formed, and there will be some subsequent intramolecular rearrangement.

Ammonia provides a special case of this type of reactivity. Experiments show that ions of the type $H_3O^+(H_2O)_n$ transfer the proton to $NH_3$, but do not always transfer water molecules to it:

$$H_3O^+(H_2O) + NH_3 \rightarrow NH_4^+ + H_2O$$
$$H_3O^+(H_2O) + NH_3 \rightarrow NH_4^+(H_2O)$$
$$H_3O^+(H_2O)_2 + NH_3 \rightarrow NH_4^+(H_2O) + H_2O$$
$$H_3O^+(H_2O)_2 + NH_3 \rightarrow NH_4^+(H_2O)_2$$

However, for the reactions of type

$$D_3O^+(D_2O)_n + NH_3 \rightarrow NH_3D^+(D_2O)_m + (n - m + 1)D_2O$$

is should be noted that there is no internal exchange between D and H. The reactions appear to take place by direct transfer of the solvated deuteron without formation of an intermediate complex. In contrast, in the corresponding rection of $D_3O^+(D_2O)_n$ with $H_2O$, the distribution of H and D in the ions produced is purely statistical, indicating formation of an intermediate complex with a long lifetime.

### 12.5.2   Solvent exchange
The reaction is

$$A^+S_n + S' \rightarrow A^+S_{n-1}S' + S$$

where S and S' are the solvating species. Consecutive reactions of this type can lead to complete replacement of the solvent, for example

$$CH_3CNH^+(H_2O)_3 + CH_3CN \rightarrow CH_3CNH^+(CH_3CN)(H_2O)_2$$
$$\downarrow CH_3CN$$
$$CH_3CNH^+(CH_3CN)_3 \xrightarrow{\quad CH_3CN \quad} CH_3CNH^+(CH_3CN)_2(H_2O)$$

**REFERENCES**

[1]  E. W. McDaniel, V. Čermák, A. Dalgarno, E. E. Ferguson and L. Friedman (eds.), *Ion–Molecule Reactions*, Wiley–Interscience, New York, 1970.
[2]  E. E. Ferguson, F. C. Fehsenfeld and A. L. Schmeltekopf, *Advan. At. Mol. Phys.*, 1969, **5**, 1.
[3]  K. G. Adams and D. Smith, *Intern. J. Mass Spectrom. Ion. Phys.*, 1976, **21**, 348.
[4]  J. B. Anderson, R. P. Andres and J. B. Fenn, *Advan. Chem. Phys.*, 1966, **10**, 275.
[5]  M. McFarland, D. L. Albritton, F. C. Fehsenfeld, E. E. Ferguson and A. L. Schmeltekopf, *J. Chem. Phys.*, 1973, **59**, 6610.
[6]  I. N. Tang, M. S. Lian and A. W. Castleman, Jr., *J. Chem. Phys.*, 1976, **65**, 4022.
[7]  P. M. Holland and A. W. Castleman, Jr., *J. Chem. Phys.*, 1908, **72**, 5984.
[8]  E. Kochanski, *J. Am. Chem. Soc.*, 1985, **107**, 7869.
[9]  H. Shinohara, N. Nishi and N. Washida, *J. Chem. Phys.*, 1985, **83**, 1939.
[10] H. W. Kroto, J. R. Heath, S. C. O'Brien, R. F. Curl and R. E. Smalley, *Nature*, 1985, **318**, 162.
[11] J. R. Heath, S. C. O'Brien, Q. Zhang, Y. Liu, R. F. Curl, H. W. Kroto, F. K. Tittel and R. E. Smalley, *J. Am. Chem. Soc.*, 1985, **107**, 7779.
[12] D. M. Cox, D. J. Trevor, K. C. Reichmann and A. Kaldor, *J. Am. Chem. Soc.*, 1986, **108**, 2547.
[13] R. G. Keesee, N. Lee and A. W. Castleman, Jr., *J. Chem. Phys.*, 1980, **73**, 2195.
[14] D. K. Bohme, J. A. Stone, R. S. Mason, R. S. Stradling and K. R. Jennings, *Intern. J. Mass Spectrom. Ion Phys.*, 1981, **37**, 283.

# 13

# Special sources

## 13.1 INTRODUCTION

For substances that are unstable, or have low volatility or high molecular weight, in most cases the mass spectrum obtained by use of electron impact ionization does not contain the molecular peak and therefore cannot be used to determine the molecular weight.

To overcome this disadvantage, special methods of ionization have been devised and satisfactory ion sources created. These include the following.

> Chemical ionization (CI) sources, either the classical low-pressure source or the newer atmospheric pressure source.
> FAB (fast atom bombardment) sources.
> Laser sources (for solid samples).
> SIMS (secondary-ionization mass spectrometry) sources.
> Field desorption sources.
> The californium source.

In almost all cases these ionization methods yield quasi-molecular peaks, $MH^+$. Cationized species such as $MNa^+$, $MK^+$, $MCa^+$ can also be observed when the appropriate metal ions are present.

## 13.2 ION SOURCES BASED ON ION–MOLECULE REACTIONS (CHEMICAL IONIZATION)

### 13.2.1 The choice of reagent gas

The choice of reagent gas for chemical ionization depends on the proton affinity $(A_p)$ of the sample molecule (M) and of the reagent gas (G). For proton transfer to take place it is necssary that $A_p(M)$ is greater than $A_p(G)$. Also, formation of fragments depends on the difference between the two proton affinities. By careful choice of the reagent gas it is possible to decrease the fragmentation and thus to obtain a very simple mass spectrum, comprising the $MH^+$ peak, adduct peaks, and only a few fragment peaks. This makes chemical ionization useful for examination of molecules which split easily under electron impact and then do not give their molecular peak in the mass spectrum. Applications of some selected reagent gases are given below.

*Methane*

The primary ions obtained from methane are $CH_4^+$, $CH_3^+$, $CH_2^+$, $CH^+$, $C^+$. If the pressure is increased, ion-molecule reactions take place between these ions and methane molecules. Under certain well defined conditions, two products predominate, $CH_5^+$ and $C_2H_5^+$ :

$$CH_4^+ + CH_4 \rightarrow CH_5^+ + CH_3$$
$$CH_3^+ + CH_4 \rightarrow C_2H_5^+ + H_2$$

If the pressure is sufficiently high, $C_3H_5^+$ is also formed.

The ions $CH_5^+$, $C_2H_5^+$ and $C_3H_5^+$ react by proton transfer and addition with the molecules (M) present in the ion source, resulting in formation of the following ions:

$$M + CH_5^+ \rightarrow MH^+ + CH_4$$
$$M + C_2H_5^+ \rightarrow MC_2H_5^+$$
$$M + C_3H^+ \rightarrow MC_3H_5^+$$

Thus the mass spectrum will contain peaks for $m/z = M+1$, $M+29$ and $M+41$. These peaks are predominant, and the spectrum is free from many of the peaks that appear in the spectrum obtained by electron impact, as shown in Fig. 13.1 for

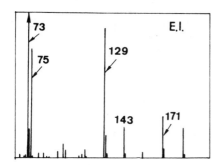 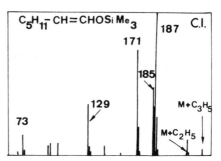

Fig. 13.1 — Comparison between spectra obtained by electron impact and chemical ionization, showing appearance of peaks for $M + C_2H_5$ and $M + C_3H_5$.

$C_5H_{11}CH = CHOSi(CH_3)_3$ ($M = 186$). The ions $M+1$, $M+29$ and $M+41$ can undergo fragmentation, but the number of these reactions is relatively small, because these ions have low internal energy.

*Isobutane*

Isobutane as reagent gas yields the ions $C_4H_{10}^+$, $C_4H_9^+$ and $C_3H_7^+$ which react with M to give ions for $M+1$, $M+43$ and $M+57$. The internal energy of these addition ions is usually lower than that of the ions formed with methane as reagent gas, so the mass spectrum is even simpler, as seen by comparison of Figs. 13.1 and 13.2.

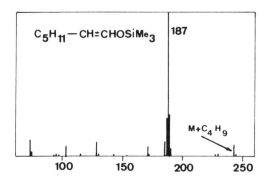

Fig. 13.2 — Chemical ionization spectrum, showing peak for M + C$_4$H$_9$.

*Ammonia*
Ammonia ia a very good reagent gas. The ions created in its ion–molecule reactions are of the type MH$^+$, MNH$_4^+$ and MNH$_3$NH$_4^+$, with masses M + 1, M + 18 and M + 35. Figure 13.3 shows a typical fragmentation pattern obtained by electron

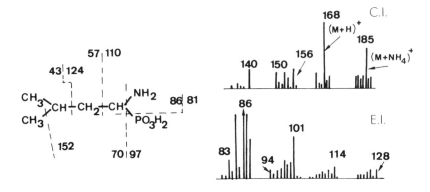

Fig. 13.3 — Comparison between electron impact ionization and chemical ionization with ammonia.

impact, and the corresponding mass spectra obtained by chemical ionization and electron impact (C$_5$H$_{14}$NO$_3$P, M = 167).

### 13.2.2 Classical chemical ionization source (CI)
The CI source is a modified electron impact (IE) source. Its features are (a) an inlet for introduction of the reagent gas, (b) very small slits, and (c) higher electron acceleration potential.

### 13.2.3   Atmospheric pressure ion source (API) [1]

An API source is a CI source designed to operate at ordinary pressures. Two factors led to its development: (a) the need to use a mass spectrometer to identify substances eluted in liquid chromatography, and (b) the use of mass spectrometry for analysis of the composition of the atmosphere. Three designs will be described.

*The TAGA source* (Fig. 13.4) [2]

This source was designed for the study of pollution in the atmosphere. The sample (air) is admitted to the source, where it is ionized, and then passes through the mass analyser. The method can be applied with or without a reagent gas (the role of which is to increase the intensity of the quasi-molecular peaks which appear after proton transfer. Ammonia, methane, etc., can be used as the reagent gas.

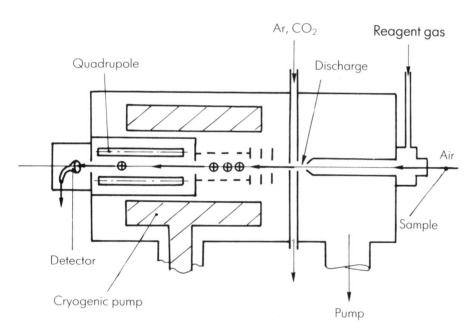

Fig. 13.4 — The TAGA instrument. (Adapted, by permission, from N. M. Reid, J. B. French, J. A. Buckley and C. C. Poon, *SCIEX Application Note 3179-P*, by courtesy of SCIEX Inc., Thornhill, Ontario, Canada.)

The sample and the reactant gas are mixed in a 22-mm diameter tube, then pass through a nozzle and an electrical discharge which ionizes the mixture. To avoid contamination of the ionization chamber by deposition on its walls, the flow-rate through the mixing tube should be fairly high (0.5–5 l./sec.). The pressure in this part of the source (the high-pressure region) is about 1 atm, and that in the analyser section is about $10^{-5}$ atm, obtained by rapid pumping with back-up from a cryogenic pump. The entrance collimator to the mass analyser has a diameter of 75 $\mu$m.

If a reactant gas is not used, the quasi-molecular ions formed are of the types

(a)  positive mode: $MH^+$, $(H_2O)_n MH^+$
(b)  negative mode: $M^-$, $(M - H)^-$

as a result of the interaction between the various molecules present and the ions $(H_2O)_n H^-$ and $(H_2O)_n O_2^-$ formed in the discharge, from the traces of water vapour present in the sample.

If a reagent gas is used, the mass spectrum contains the quasi-molecular peak for $(M + H)^+$ and the adduct peaks, the composition of which depends on the reactant gas, e.g. $(M + NH_4)^+$ for $NH_3$, $(M + C_2H_5)^+$ etc. for $CH_4$.

In both cases (with or without reagent gas), the mass spectrum contains peaks arising from fragmentation. The persistence of clusters, their passage into the analysis region, and blockage of the collimator slit by deposition of particles, are all prevented by the introduction of a sheath gas (Ar, $CO_2$, etc.) to create a barrier between the ionization and analysis regions.

The whole analysis unit is called a TAGA (trace atmospheric gas analyser) spectrometer. It comprises an atmospheric pressure source, a quadrupole analyser and a detector.

Typical applications are as follows.

(a)  Determination of pollutants in the air, atmospheric pollution checks, and establishing the spatial distribution of pollutants.
(b)  Analysis of gases exhaled by subjects exposed to toxic substances.
(c)  Analysis of organic substances extracted from the ground (such as landfill gases); in this case a preconcentration unit is combined with the source, as shown in Fig. 13.5, which allows for accumulation of the analyte on the absorber, and its subsequent desorption into the ionization region.

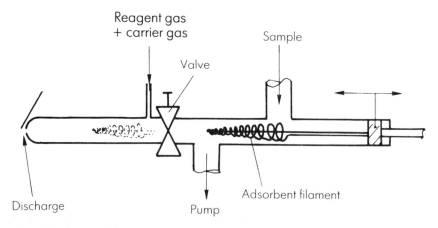

Fig. 13.5 — Apparatus for preconcentration by adsorption. (Adapted, by permission, from B. A. Thomson, T. Sakuma, J. Fulford, D. A. Lane, N. M. Reid and J. B. French, *SCIEX Application Note 3079-P*, by courtesy of SCIEX Inc., Thornhill, Ontario, Canada.)

An example is the use of the TAGA source to establish the atmospheric distribution of benzothiazole in the vicinity of a factory, as a function of the prevailing wind conditions (Fig. 13.6).

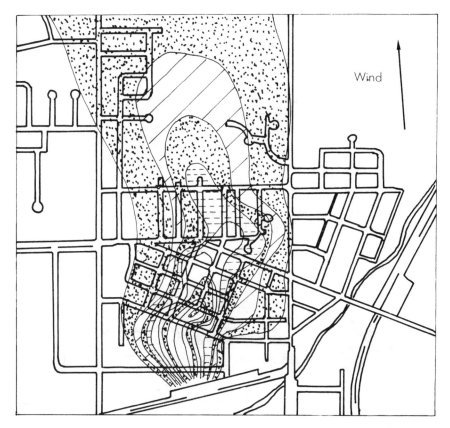

Fig. 13.6 — Distribution of benzothiazole. (Adapted, by permission, from N. M. Reid, J. B. French, J. A. Buckley and C. C. Poon, *SCIEX Application Note 3179-P*, by courtesy of SCIEX Inc., Thornhill, Ontario, Canada.)

*The Horning source* [3]

This source (shown diagramatically in Fig. 13.7) is intended for use in liquid chromatography coupled with mass spectrometry. The liquid sample is injected into a heated vaporizer tube, and the vapour from it is carried by an inert carrier gas into the reaction chamber, where it is ionized by interaction with electrons from a radioactive $^{63}$Ni foil. The ions and neutral molecules enter the mass analyser through a 25-$\mu$m diameter hole. Much greater sensitivity and dynamic range can be achieved by use of a corona discharge (at 500–2000 V) instead of the $^{63}$Ni for the ionization, and with iso-octane as the eluent in the chromatography [4]. $(M + H)^+$ ions are obtained. The two types of source have been compared.

*The nebulizer source* [5–7]

The effluent from a liquid microchromatograph is introduced into the nebulizer through a coaxial stainless-steel capillary (A in Fig. 13.8) in the nebulizer tube (C) into which the nebulizer gas, helium, is fed through a needle (B). The nebulizer tip

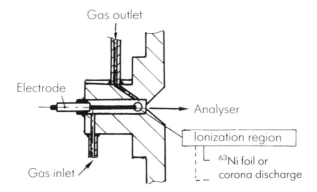

Fig. 13.7 — Schematic diagram of a Horning-type source. The electrode is needed only for the corona-discharge type of source.

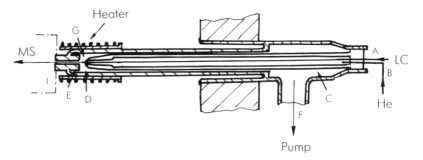

Fig. 13.8 — Nebulizer source. (Adapted, by permission, from S. Tsuge, Y. Hirata and T. Takeuchi, *Anal. Chem.*, 1979, **51**, 166. Copyright 1979, American Chemical Society, Washington, D.C.)

(D) is aligned with the entrance capillary (E) into the chemical ionization chamber (I) and the distance between the two is adjusted so that the pressure in the CI chamber is about 1 torr. The nebulization chamber (G) and the entrance capillary (E) can be heated to prevent adsorption of low-volatility compounds on the capillary walls. The pressure in the nebulization chamber is reduced by pumping through F. The effluent from the chromatograph is drawn into the nebulizer, where it is converted into a spray of very fine droplets of mobile phase or of mobile phase plus analyte, depending on the elution pattern, and the solvent is evaporated either under the reduced pressure or by heating (or both), and a fraction of the effluent molecules will pass into the CI chamber. The remainder will be pumped away, and the lighter molecules may be preferentially removed in this way. The fraction entering the CI chamber then undergoes chemical ionization, the reactive gas being either the mobile phase solvent or a selected reactant gas admitted directly to the CI chamber. A core wire placed coaxially in capillary A gives better reproducibility and aids in cleaning.

*Electrospray*
Production of a mist of very fine droplets by passage of a liquid through a capillary carrying a high electrostatic charge dates back to 1917 [8] but was not used in mass spectrometry until the pioneer work of Dole *et al.* on polymer solutions [9]. It is now used as an LC/MS interface and is useful for high molecular-weight substances, e.g. gramicidin S. The sample is passed through a metal capillary in a high potential field where it is converted into a fine spray of charged droplets, from which the solvent rapidly evaporates. The solute ions are then transferred into the mass analyser [10].

*Thermospray* [11]
This method requires the sample to contain an electrolyte that can be volatilized, such as ammonium acetate. The liquid sample is passed through an electrically heated stainless-steel capillary tube into a heated evacuated chamber. The liquid leaving the capillary is in the form of a supersonic jet of very fine droplets, from which the solvent rapidly evaporates, leaving the analyte molecules in a sheath of electrolyte. The electrostatic field generated by the electrolyte ions rapidly increases as the droplets decrease in size, and results in a soft ionization of the analyte. This ionization process does not always work, however, and may be backed up by use of an electric discharge or an external source of electrons. The ions are admitted into the mass spectrometer through a sampling cone. The apparatus is shown schematically in Fig. 13.9.

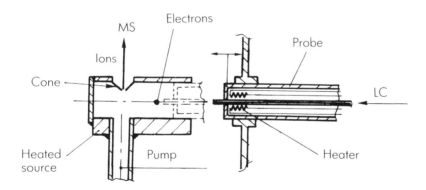

Fig. 13.9 — A thermospray. (Adapted, by permission, from M. L. Vestal, *Eur. Spectrosc. News*, 1985, No. 63, 22. Copyright, 1985, John Wiley and Sons Inc., New York.)

*MAGIC*
This device (the monodisperse aerosol generator for introduction of liquid chromatography effluents) provides an alternative to thermospray, for use with eluents that do not contain an electrolyte. A specially designed nebulizer produces an aerosol having a very narrow range of droplet sizes; the droplets are desolvated at atmospheric pressure and passed through a beam separator into the mass spectrometer, where electron impact or chemical ionization can be applied [12].

### 13.3   FAST ATOM BOMBARDMENT (FAB) [13–14]

The precursor of this technique was the method of solid-surface analysis described by Devienne and Roustan [15]. The components of an argon FAB source for the formation of the primary atom beam are a cold-cathode discharge source producing argon ions with an energy between 2 and 10 keV, and a system for focusing the $Ar^+$ ions in a collision chamber containing argon at $10^{-4}$–$10^{-3}$ torr pressure (Fig. 13.10). In the collision chamber resonant charge-transfer takes place between the argon ions and argon atoms. In this process the momentum of the reacting argon ions is practically completely conserved, so the argon atoms produced by the charge transfer have the same velocity as their parent ions. The gas leaving the chamber will still contain some argon ions because of incompleteness of the charge-transfer reaction, but these can be removed by means of a pair of deflector plates, and the final beam will consist practically entirely of argon atoms.

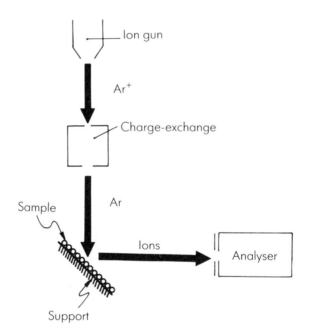

Fig. 13.10 — A fast-atom bombardment system.

### 13.3.1   The sample
The substance to be analysed is placed on a metal sample holder (generally made of stainless steel or copper) fixed to a direct introduction probe which is introduced into the source through an air-lock.

### 13.3.2   Ionization of the sample
The atom beam takes the place of the electron beam in an electron-impact ionization

source, bombarding the sample and forming ions from it. These ions are then extracted, accelerated, focused and analysed by the mass analyser. The extraction is done by means of an electric field at the potential that would be used for the electron beam in electron-impact ionization. The metal sample holder is at the same potential as the repeller plate (cf. Chapter 3). The other plates are kept at the working potential used for electron-impact ionization. The ionization conditions can be varied by changing the energy of the argon atoms (in the range 2–10 keV) and the atom flux ($10^{10} - 10^{11}$ atoms. cm$^{-2}$. sec$^{-1}$).

### 13.3.3   Mass spectra
The mass spectrum obtained by FAB contains peaks for quasi-molecular ions and ion fragments characteristic of the analyte, together with those for cationized products, e.g. $(M + H)^+$, $(M - H)^+$, $(M + Na)^+$, $(M - H + Na)^+$.

### 13.3.4   FAB with Cs, Xe etc.
FAB systems using caesium, xenon or other atoms for excitation work on the same principle, the sample being bombarded with a fast stream of atoms.

### 13.3.5   The capillaritron (mini-FAB)
The apparatus described above has two ports — one for admission of the atom beam, the other for introduction of the sample. A new type of apparatus needs only one port: the sample is placed on a sample holder carried by a probe in the path of the argon or caesium beam.

### 13.3.6   Characteristics of the FAB method
These are as follows.

(1) The sample is ionized at low temperature (i.e. without heating).
(2) Quasi-molecular ions and characteristic fragments are produced.
(3) Both positive and negative ions can be formed.
(4) The sample has a long lifetime, making it possible to work for an extended period with the same sample, which assists in (a) establishing a fragmentation pattern by using the methods for study of metastables (Chapter 10), and (b) establishing the atomic composition of the ions by use of high resolution.
(5) High ion intensities and relatively stable peak heights are obtained.

A typical example of application of the technique is to gramicidin S, as shown in Fig. 13.11, which gives the mass spectrum and the fragmentation pattern deduced from it.

## 13.4   LASER ION SOURCES [16]

### 13.4.1   Introduction
Lasers can be used in several ways in mass spectrometry.

(1) For ionization of a sample in the gas phase.
(2) To vaporize a sample on a sample holder. In this case electron impact or a second laser is used for the ionization, and the sample holder is introduced directly into the desorption laser beam.

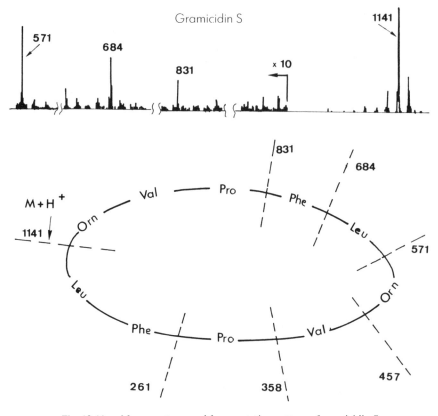

Fig. 13.11 — Mass spectrum and fragmentation pattern of gramicidin S.

(3) For desorption of a substance from a filament, as in field desorption (FD) ionization, where the desorbed molecules are ionized by an electric field.

(4) For simultaneous desorption and ionization of molecules. The substance is placed on a sample holder which is introduced directly into the ion source.

(5) For simultaneous desorption and ionization in the study of the spatial distribution of anions and cations. A microsample is placed in the ion source and scanned by the laser beam.

(6) In single-photon (at one wavelength) or multi-photon (at one, two or three wavelengths) mode.

Methods (3) and (4) are used for organic molecules. Method (2) has been used for inorganic compounds, and method (5) for various anions and cations. The main applications are to inorganic substances.

The lasers used for organic molecules have wavelengths ranging from the infrared to the far ultraviolet.

Two modes of operation have been explored: (1) the normal mode; (2) the Q-switched (pulsed) mode. With the latter mode very high energy densities can be obtained (typically $10^9$–$10^{11}$ W/cm$^2$) in a very short time interval, with a relatively low repetition rate (e.g. less than 100 Hz). The effect of the laser beam depends on its wavelength and the pulse duration and repetition rate.

### 13.4.2  Examples of types of laser used
*The CO₂ laser*
The wavelength is 1.06 $\mu$m, the energy density (after focusing) from 30 W/cm² to several kW/cm² for continuous running, and 1 MW/cm² in the pulsed mode.

*The Nd laser*
The wavelength is 1.06 $\mu$m, and the energy density 100 MW/cm² in the pulsed mode.

*The Nd–YAG laser*
Equipped with a frequency multiplier ( × 2,  × 3) this laser gives wavelengths of 353 and 265 nm respectively, and the energy density ranges from $10^4$ to $10^5$ MW/cm².

*The ruby laser*
This laser can be used in either normal or Q-switched mode, with or without frequency multiplication, and gives energy densities of 10–100 MW/cm².

### 13.4.3  Samples
The sample can be in the form of a solution (in ethanol, methanol, water, etc.) and is placed on the sample holder with a syringe, or deposited by sputtering. The sample holder is a metal grid (usually stainless steel) and is positioned in the source by means of a micromanipulator (Fig. 13.12).

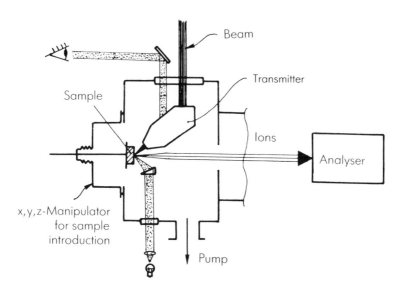

Fig. 13.12 — Arrangement for laser desorption and ionization.

### 13.4.4  Characteristics of laser desorption
(1) The ions have quite a wide energy range.
(2) The depth of penetration by the laser beam depends on the beam intensity. A fairly intense beam forms a plasma of neutral particles, anions and cations on the surface.

(3) The desorption time is relatively short, so it is necessary to scan the laser beam over the surface of the sample.

### 13.4.5 The ion image
For a heterogeneous sample, laser desorption produces an ion image which represents the spatial distribution of the chemical elements, molecules or functional groups present in the sample. Successive application of the positive and negative ion modes results in cation and anion images, which give complementary information and facilitate identification of the unknown and derived species. The ion images are obtained by scanning the laser over the surface of the sample and recording the successive spectra obtained. The spatial resolution is better than 1 $\mu$m.

Research has been done on microsamples of biological tissues (biological cross-sections) and on inorganic samples. The preparation of the biological samples must not change the distribution of the water-soluble chemical species (e.g. cations). For this reason the sample is freeze-dried, preserved, and cut to give sections with a thickness of 0.3–1 $\mu$m. Interesting results have been obtained on the distribution of metals in muscle and uterine tissue (Fe), and biopsy of the retina (Na, Mg, Ca, Br) and brain tissue (Li).

### 13.4.6 The ions formed
The positive ion mass spectrum contains peaks for the quasi-molecular cations such as $(M + H)^+$, $(M + Na)^+$, $(M - H + Na)^+$ and various fragment ions [17,18].

### 13.5 THE CALIFORNIUM ($^{252}$Cf) SOURCE [19–22]

### 13.5.1 Source and samples
Californium-252 is an isotope with a half-life of 2.65 years. It decomposes by two paths: by $\alpha$-particle decay (96.9%) and by spontaneous fission (3.1%). The fission route produces pairs of charged particles with unequal masses and energies, which travel away from each other in almost opposite directions, as shown schematically in Fig. 13.13. The largest disparity in mass is that for the production of $^{142}$Ba and $^{106}$Tc:

$$^{252}\text{Cf} \rightarrow \begin{matrix} ^{142}\text{Ba} \\ (79 \text{ MeV}) \end{matrix} + \begin{matrix} ^{106}\text{Tc} \\ (104 \text{ MeV}) \end{matrix}$$

There are numerous other combinations of fission fragments.

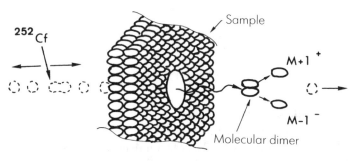

Fig. 13.13—Californium-252 source. (Adapted, by permission, from R. D. Macfarlane and D. F. Torgerson, *Science*, 1976, **191**, 920. Copyright 1976, American Association for Advancement of Science, New York.)

The sample, in the form of a solid film, is placed on a metal foil opposite the $^{252}$Cf source. The fission produces two particles, and the one which moves towards the sample produces collisional desorption and ionization of the molecules at the point of impact. In effect, the sample is locally heated and the substance is volatilized and ionized. The second particle (travelling in the opposite direction) is used to trigger a clock and start acquisition of the signal by the collector.

The californium source is prepared by sputtering. The $^{252}$Cf is deposited on a nickel foil (surface area 1.2 cm$^2$, thickness 1 $\mu$m) and covered with gold leaf to protect it from contamination from the surroundings. The foil sandwich is then fixed on a circular aluminium support and placed near the foil carrying the sample for analysis (Fig. 13.14). The ions emitted from the sample are analysed by a time-of-flight analyser.

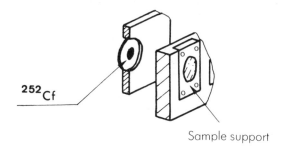

$^{252}$Cf

Sample support

Fig. 13.14 — Sample and source assembly for the $^{252}$Cf source.

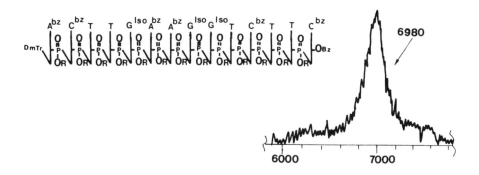

Fig. 13.15 — Example of mass spectrum obtained with a californium source. (Adapted, by permission, from C. J. MacNeal, *Anal. Chem.,* 1982, **54**, 43A. Copyright 1984, American Chemical Society, Washington D.C.)

### 13.5.2  The ions formed

Quasi-molecular ions and fragment ions are observed in the mass spectrum; a typical example is shown in Fig. 13.15.

**REFERENCES**

[1] E. C. Horning, M. G. Horning, D. I. Carroll, I. Didzic and R. N. Stillwell, *Anal. Chem.*, 1973, **45**, 936.

[2] N. M. Reid, J. A. Buckley, C. C. Poon and J. B. French, *Adv. Mass Spectrom.*, 1980, **8B**, 1843.

[3] D. I. Carroll, I. Dzidic, R. N. Stillwell, K. D. Haegele, M. G. Horning and E. C. Horning, *Anal. Chem.*, 1974, **46**, 706.

[4] D. I. Carroll, I. Dzidic, R. N. Stillwell, K. D. Haegele and E. C. Horning, *Anal. Chem.*, 1975, **47**, 2369.

[5] S. Tsuge, Y. Hirata and T. Takeuchi, *Anal. Chem.*, 1979, **51**, 166.

[6] Y. Hirata, T. Takeuchi, S. Tsugi and Y. Yoshida, *Org. Mass Spectrom.*, 1979, **14**, 126.

[7] Y. Yoshida, H. Yoshida, S. Tsuge and T. Takeuchi, *HRC & CC*, 1980, **3**, 16.

[8] J. Zeleny, *Phys. Rev. Ser.*2, 1917, **10**, 1.

[9] M. Dole, L. L. Mack, R. L. Hines, R. C. Morley, L. D. Ferguson and M. B. Alice, *J. Chem. Phys.*, 1968, **49**, 2240.

[10] C. M. Whitehouse, R. N. Dreyer, M. Yamashita and J. B. Fenn, *Anal. Chem.*, 1985, **57**, 675.

[11] M. L. Vestal, *Eur. Spectrosc. News*, 1985, No. 63, 22.

[12] R. C. Willoughby and R. F. Browner, *Anal. Chem.*, 1984, **56**, 2626.

[13] D. S. Surman and J. C. Vickerman, *J. Chem. Soc. Chem. Commun.*, 1981, 324.

[14] M. Barber, R. S. Bordoli, R. D. Sedgwick and A. N. Tyler, *J. Chem. Soc. Chem. Commun.*, 1981, 325.

[15] F. M. Devienne and J.-C. Roustan, *C. R. Acad. Sci. Paris*, 1976, **283B**, 397.

[16] R. J. Cotter, *Anal. Chem.*, 1984, **56**, 485A.

[17] R. Stoll and F. W. Röllgen, *Z. Naturforsch.*, 1982, **37a**, 9.

[18] G. J. Q. van der Peyl, J. Haverkamp and P. G. Kistemaker, *Intern. J. Mass Spectrom. Ion Phys.*, 1982, 42, 125.

[19] R. D. MacFarlane and D. F. Torgenson, *Science*, 1976, **191**, 920.

[20] R. D. MacFarlane and D. F. Torgenson, *Intern. J. Mass Spectrom. Ion Phys.*, 1976, **21**, 81.

[21] R. D. MacFarlane, *Biomed. Mass Spectrom.*, 1981, **8**, 449.

[22] C. J. McNeal, *Anal. Chem.*, 1982, **54**, 43A.

# 14

# Applications of mass spectrometry in atmospheric investigations

## 14.1 INTRODUCTION

At the end of the 19th century the hypothesis was advanced that the atmosphere contains an electrically charged layer. This hypothesis was revived to explain the propagation of radio waves, especially their reflection at an altitude of about 50 km.

Figure 14.1 shows the structure of the atmosphere. The ionosphere can be divided into several layers, which have particularly interesting characteristics.

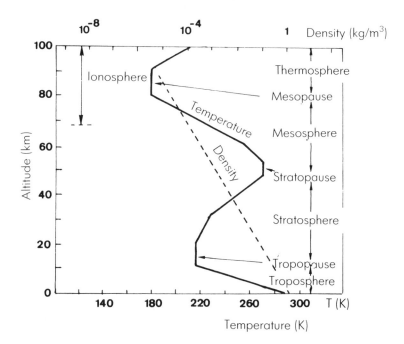

Fig. 14.1 — Layers in the atmosphere, and the atmospheric temperature distribution.

Layer 'D' lies at an altitude of between about 60 and 85 km, and is mainly due to the ionization of NO by Lyman $\alpha$ radiation of wavelength 121.6 nm. At altitudes below 70 km high-energy cosmic radiation contributes to the charge in the layer by ionizing $N_2$ and $O_2$. This layer is highly absorbant for radio waves.

Layer 'E' is at an altitude of about 85–130 km, and its ionization is mainly due to low-energy X-radiation, but there is also a contribution from the Lyman $\beta$ radiation. The primary ions produced include $O_2^+$, $NO^+$ and various ions ($Fe^+$, $Ca^+$, $Mg^+$, $Si^+$, etc.) that are meteoric in origin. The $Fe^+$ and $Ca^+$ ions are present in a permanent layer at between 85 and 100 km, and also in fine layers (1–3 km thick) between 100 and 105 km. These metal ions are formed by charge-transfer reactions with $O_2^+$ and $NO^+$, by photionization, or by direct ionization on vaporization. They have a relatively long lifetime. Below 90 km their concentration rapidly decreases. The presence of a layer of $Na^+$ and the presence of $K^+$ and $Li^+$ may also be noted.

Layer 'F', located at above 130 km, consists of two sub-layers, $F_1$ and $F_2$. The ionizing factor is extremely short wavelength ultraviolet radiation (9–19 nm) and the species ionized are oxygen atoms and nitrogen molecules. In layer $F_2$ the charge concentration also depends on transport by molecular diffusion. It also varies during the day, being highest at noon local time.

Above the F layer the predominant ions are $He^+$ and $H^+$. In this region the ions are distributed according to the earth's magnetic field, and the notion of horizontal layers becomes meaningless.

The main ionization processes in the troposphere are due to radiation from the sun, cosmic radiation, and the diffusion of ions from the stratosphere ($N^+$, $O^+$, $O_2^+$, $N_2^+$, $Ar^+$, $H_2O^+$, $OH^+$). Figure 14.2 shows the penetration of the atmosphere by

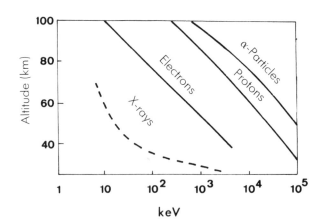

Fig. 14.2 — Penetration of the atmosphere by different particles, as a function of their energy. (Reproduced by permission, from G. Brasseur and S. Solomon, *Aeronomy of the Middle Atmosphere*. Copyright 1984, Reidel, Dordrecht.)

various species and X-rays. The more predominant secondary ions formed by ion–molecule reactions include $O_2^+$, $O_4^+$, $N_3^+$, $N_4^+$, $ArH^+$, $H_3O^+$, $NO^+$, $O^+(H_2O\text{-}_n)$, $NO^+(H_2O)_n$, $NO_2^+(H_2P)_n$. The rates of ionization by cosmic rays are shown in Fig. 14.3, and the contributions of permanent ionization sources to production of ions in the middle atmosphere are given in Table 14.1. Figure 14.4 shows the principal reactions leading to formation of $NO^+$ in layer E of the ionosphere.

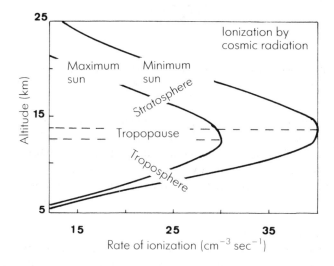

Fig. 14.3 — Rate of ionization of air by cosmic radiation in the lower stratosphere and the troposphere. (Adapted, by permission, from G. Brasseur and M. Nicolet, *Planet. Space Sci.*, 1973, **21**, 939. Copyright 1973, Pergamon Press Ltd. Oxford.)

**Table 1** — Ionization in the middle atmosphere: contribution of various factors.

| Source | Flux (erg. cm$^{-2}$. sec$^{-1}$) |
|---|---|
| Direct solar $\alpha$-Lyman | 6 |
| Scattered solar $\alpha$-Lyman (night-time) | $6 \times 10^{-3}$–$6 \times 10^{-2}$ |
| Galactic cosmic rays | $10^{-3}$–$10^{-2}$ |
| Cosmic X-ray, $\lambda$ 0.1–1 nm | $4 \times 10^{-9}$ |
| Solar X-rays, $\lambda$ 1–10 nm | 0.1–1 |
| Magnetosphere electrons, | |
| auroral zone | 0.1–1 |
| middle latitudes | $10^{-4}$–$10^{-3}$ |

## 14.2  STUDY OF PROCESSES OCCURRING IN THE ATMOSPHERE

Mass spectrometry has been used in two ways in the study of atmospheric processes: for *in situ* measurements and for simulation of the processes. The aim of such

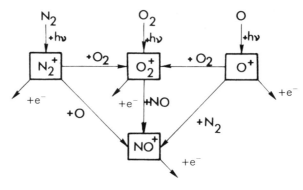

Fig. 14.4 — Reactions in zone E of the ionosphere. (Reproduced by permission, from G. Brasseur and S. Solomon, *Aeronomy of the Middle Atmosphere*. Copyright 1984, Reidel, Dordrecht.)

research is to identify the ions present, establish their reaction behaviour, and describe their temporal and spatial evolution. The results obtained have been used to propose models for the structure and composition of the ionosphere, stratosphere and troposphere. Mass spectrometry has thus contributed by means of direct observation and laboratory experiments to the construction of theoretical models of the atmosphere.

### 14.2.1   *In situ* measurements

The mass spectrometers used for *in situ* measurements in the atmosphere are remote-controlled quadrupole instruments [1], and are used to detect neutral species and positive and negative ions in the stratosphere (20–30 km), region D (50–85 km) and the lower ionosphere (about 70–120 km). The ions are detected with a channeltron, and the detection limits are set by the background noise of the channeltron working in the pulse-counting mode. The vacuum needed is maintained with a cryogenic pump cooled with liquid nitrogen. Figure 14.5 shows schematically the mass

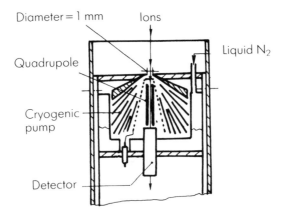

Fig. 14.5 — Schematic diagram of a mass spectrometer mounted in a rocket for space research. (Adapted, by permission, from R. S. Narcisi and A. D. Bailey, *J. Geophys. Res.*, 1965, **70**, 3687. Copyright 1965, American Geophysical Union, Washington, D.C.)

spectrometer used in an early survey of the D and lower E layers [1]. The sampling unit was separated from the analyser unit by a diaphragm with a small orifice, and an electron multiplier tube was used as the detector. Figure 14.6 shows the principle of

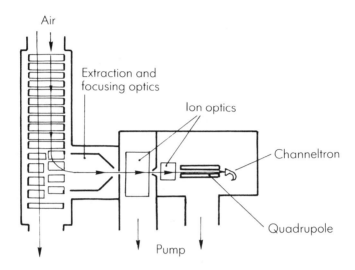

Fig. 14.6 — Principle of an apparatus for mass analysis of ions in the troposphere. (Adapted, by permission, from F. L. Eisele, *Intern. J. Mass Spectrom. Ion Processes*, 1983, **54**, 119. Copyright 1983, Elsevier, Amsterdam.)

the apparatus used for analysis of the troposphere [2]. In this, the air sample passed through a specially designed sampling tube, and a test sample was admitted to the mas spectrometer through a collimating system. The ions present were then directly detected with the quadrupole analyser (channeltron system). Neutrals were identified by ionization of the sample.

### 14.2.2   Results obtained *in situ*

Measurements taken at between 65 and 90 km altitude (Fig. 14.7) show the presence of $H^+(H_2O)$ clusters with $m/z = 19$ and 37 ($H_5O_2^+$), and $O_2^+$ ($m/z = 32$) and $NO^+$ ($m/z = 30$) ions.

At a height of about 80 km the dominant species is $H^+(H_2O)_n$, where the value of $n$ depends on altitude and on physical conditions such as temperature. Ions with $n = 2$ ($m/z = 37$) and $n = 4$ ($m/z = 73$) are the most abundant of these species, but at low temperatures, near the mesopause, ions with $n = 8$ or 9 can be observed.

Ions in the stratosphere were first detected by mass spectrometry in 1977 [3] and it was shown that at above 45 km the main species are protonated water clusters of the type $H_3O^+(H_2O)_n$, whereas below this height the dominant species are non-protonated hydrates with $m/z = 29 \pm 2$, $42 \pm 2$, $60 \pm 2$ and $80 \pm 2$. Later [4], ions with $m/z = 96 \pm 2$ were found in addition to these ions. Other observations made in 1978

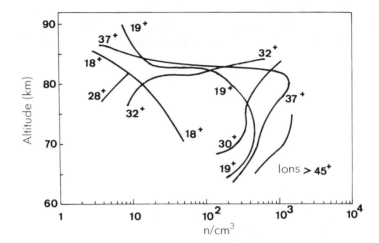

Fig. 14.7 — Positive ions detected at altitudes between 65 and 90 km by an *in situ* spectrometer. (Reproduced by permission, from R. S. Narcisi and A. D. Bailey, *J. Geophys. Res.*, 1965, **70**, 3687. Copyright 1965, American Geophysical Union, Washington, D.C.)

[4,5] and 1980 [6] showed the presence of ions of the type $H^+X_n(H_2O)_m$, where X has a mass of 41; X is now regarded as almost certainly being $CH_3CN$ [7].

Ions with $m/z = 239$, 240 and 241 have been found in the troposphere.

### 14.2.3   Simulation of atmospheric processes
Equipment of the type described for the study of ion clusters and ion–molecule reactions (Chapters 11 and 12) is used. Most of the results have been obtained with a flowing afterglow unit, the studies by Ferguson and co-workers (e.g. [8–10]) being particularly important.

### 14.2.4   Ion–molecule reactions in the atmosphere
The mechanisms postulated for ion formation in zone D and in the stratosphere are shown in Figs. 14.8 and 14.9. These schemes are based on interpretation of *in situ* measurements and the results of laboratory experiments, and are in accordance with both sets of information.

Laboratory experiments have shown that $CH_4$, $CO_2$ and $O_3$, which occur at ppm v/v concentrations in the atmosphere, are not included in the diagrams, since they do not play a direct role in the production of the $H_3O^+(H_2O)_n$ clusters. The species HCHO, $HNO_3$ and $N_2O_5$, which occur at parts in $10^9$ v/v concentrations are likewise excluded, though they can alter the reaction chain. Methanol and ammonia react irreversibly with the protonated water clusters, e.g.

$$H^+(H_2O)_n + NH_3 \rightarrow NH_4^+(H_2O)_m + (n - m)H_2O$$

Data are available for the gas-phase reactions of protonated water clusters with a

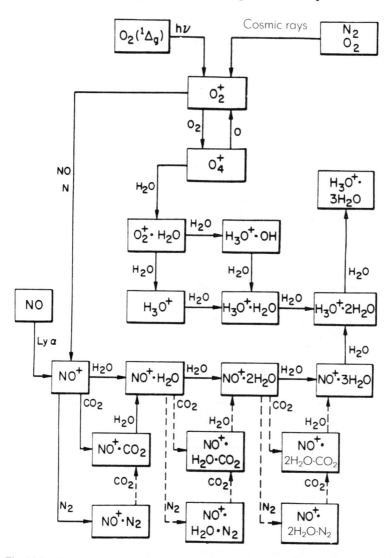

Fig. 14.8 — Ion–molecule reactions occurring in zone D. (Reproduced by permission, from E. E. Ferguson, in *Kinetics of Ion–Molecule Reactions*, P. Ausloos (ed.), Plenum Press, New York, 1979.)

wide variety of molecules in the troposphere, such as $CFCl_3$, $H_2SO_4$, $H_2SO_3$, $CH_3CHO$, $H_2S$, $HCOOH$, $CH_3COOH$, $(CH_3)CO$, $(CH_3)_2O$, $(C_2H_5)_2S$, $C_6H_5OH$, $C_6H_5NH_2$, $C_6H_5NO_2$. The generalized reaction is

$$H_3O^+(H_2O)_n + X \rightarrow XH^+(H_2O)_m + (n - m + 1)\ H_2O$$

As might be expected, the negative ion chemistry of the atmosphere is also very

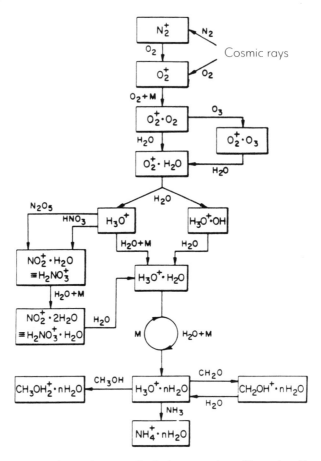

Fig. 14.9 — Ion–molecule reactions occurring in the stratosphere. (Reproduced by permission, from E. E. Ferguson, in *Kinetics of Ion–Molecule Reactions*, P. Ausloos (ed.), Plenum Press, New York, 1979.)

interesting and complex [9], offering even more numerous possibilities for interaction with the positive ions present. A good account is given by Brasseur and Solomon [11].

**REFERENCES**

[1] R. S. Narcisi and A. D. Bailey, *J. Geophys. Res.*, 1965, **70**, 3687.
[2] F. L. Eisele, *Intern. J. Mass Spectrom. Ion Processes*, 1983, **54**, 119.
[3] F. Arnold, D. Krankowsky and H. K. Marien, *Nature*, 1977, **267**, 30.
[4] E. Arijs, J. Ingels and D. Nevejans, *Nature*, 1978, **271**, 642.
[5] F. Arnold, H. Boehringer and G. Henschen, *Geophys. Res. Lett.*, 1978, **5**, 653.
[6] E. Arijs, D. Nevejans and J. Ingels, *Nature*, 1980, **288**, 684.

[7]  H. Boehringer and F. Arnold, *Nature*, 1981, **290**, 321.
[8]  F. C. Fehsenfeld and E. E. Ferguson, *J. Chem. Phys.*, 1974, **61**, 3181.
[9]  F. C. Fehsenfeld, I. Dotan, D. L. Albritton, C. J. Howard and E. E. Ferguson, *J. Geophys. Res.*, 1978, **83**, 133.
[10]  F. C. Fehsenfeld and E. E. Ferguson, *J. Geophys. Res.*, 1969, **74**, 2217.
[11]  G. Brasseur and S. Solomon, *Aeronomy of the Middle Atmosphere*, Reidel, Dordrecht, 1984.

# 15

## Theory of spectra

### 15.1 DIATOMIC MOLECULES

#### 15.1.1 The Morse curve

In a diatomic molecule AB the two atoms oscillate about their equilibrium positions by stretching and contraction of the bond between them, as shown in Fig. 15.1.

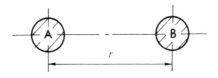

Fig. 15.1 — The atoms vibrate about their equilibrium positions, along the interatomic axis.

If such a system behaved as a simple harmonic oscillator, the potential energy of the system would be a parabolic function of the internuclear distance $r$. In reality, however, this is not the case. There will be a minimum in the potential energy curve (Fig. 15.2) when the two atoms are at their equilibrium positions ($r = r_0$, the equilibrium bond length), and for the region in which $r$ is not much different from $r_0$ the potential energy curve is indeed parabolic, but when the two atoms become very close to each other the potential energy rises more rapidly than expected, because of the physical repulsion between them. On the other hand, when the two atoms are moving apart, their momentum (if their kinetic energy $E$ is large enough) makes their displacement from each other larger than expected and the vibration is anharmonic. If the total vibrational energy $E_v$ is high enough, the distance between the two atoms may become large enough for the bond between them to be broken ($c$ in Fig. 15.2).

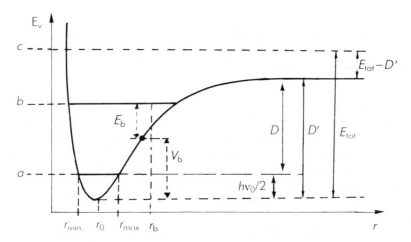

Fig. 15.2 — The Morse curve. Level $a$ corresponds to the fundamental vibration $v_0$. For level $b$, when $r = r_b$ the vibrational energy is divided between the kinetic energy $E_b$ and the potential energy $V_b$. Level $c$ corresponds to an energy input greater than that needed for dissociation, so the fragments acquire translational kinetic energy. For $D'$ see text.

This will occur when the potential energy $V$ of the system is equal to or greater than the dissociation energy $D$. An exact mathematical description of the system is difficult, but the Morse curve (shown in Fig. 15.2) is a good approximation; it is described by the expression

$$V = D'(1 - \exp[-\beta(r - r_0)])^2$$

where $D'$ is the potential energy measured from the minimum of the Morse curve (which is taken as zero) to the energy corresponding to dissociation, and $\beta$ is a constant related to $D'$, the fundamental vibrational frequency $v_0$ and the reduced mass of the molecule. The vibrational energy levels of the molecule are quantized, with energy levels given by

$$E_n = (n + \tfrac{1}{2})hv_0$$

where $n$ is the vibrational quantum number, taking the values $0, 1, 2\ldots$, and $h$ is Planck's constant. Thus at its lowest vibrational level the molecule still has an energy of $hv_0/2$ (the zero-point energy), $a$ in Fig. 15.2. Hence $D'$ is related to the bond dissociation energy $D$ (the energy required to dissociate the molecule at absolute zero) by

$$D' = D + hv_0/2$$

The quantized energy levels are illustrated in Fig. 15.3. If the energy imparted to the system is greater than the dissociation energy, the excess is transformed into translational kinetic energy of the fragments (c in Fig. 15.2).

### 15.1.2   Excited electronic states

For each electronic state of the molecule there are corresponding potential energy curves, which can take two forms: the Morse curve as in a and b in Fig. 15.4, or a curve without a minimum, as in c and d in Fig. 15.4. If a molecule is excited to a state giving the second type of curve, it will dissociate. This second form of curve can be associated with the promotion of bonding electrons into non-bonding or anti-bonding orbitals, resulting in repulsion between the two atoms, and dissociation of the molecule.

The dissociation energy will depend on the electronic state of the molecule, and the dissociation products may be formed in an excited state if the input excitation energy is high enough.

### 15.1.3   Intersection of potential energy curves; predissociation

If the potential energy curve for an excited electronic state has no minimum, but crosses the Morse curve for a lower energy state (A in Fig. 15.5) the molecule can dissociate at an energy lower than that of the asymptote of the Morse curve, since there can be a 'cross-over' from one curve to the other at the intersection, as a result of reorganization of the electrons (called an internal conversion). This phenomenon is called predissociation.

In contrast, if two Morse-type curves intersect (B in Fig. 15.5) there will be a region in which both sets of vibrational states can occur, but there will also be predissociation if the dissociation energy for the excited state is lower than that for the lower electronic energy state.

### 15.1.4   The Franck–Condon principle

The electronic absorption bands in the spectra of liquids can sometimes be resolved to show the presence of vibrational bands. This is accounted for by the Franck–Condon principle, which is based on the fact that electronic transitions take only about $10^{-15}$ sec, whereas a vibration takes about $10^{-12}$ sec, the result being that the positions of the nuclei remain practically unchanged during an electronic transition. Generally, an electronic excitation results in an increased bond length, and the minimum of the Morse curve for an excited state will then be at a larger $r$ value than that for the ground state. It follows that for a given internuclear distance there are several possible transitions from one vibrational level in the electronic ground state to different vibrational levels in the excited state.

Since such transitions can easily be identified on the Morse curve plots by drawing a vertical line at the appropriate $r$ value, they are called **vertical transitions**. In contrast, a transition from the fundamental vibrational state ($n = 0$) in the electronic ground state to the corresponding level in the excited electronic state is called an **adiabatic transition**, and involves a change in the internuclear distance.

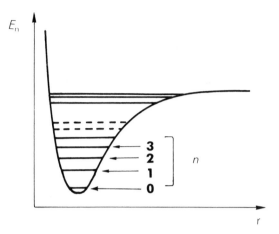

Fig. 15.3 — Quantized vibrational energy levels.

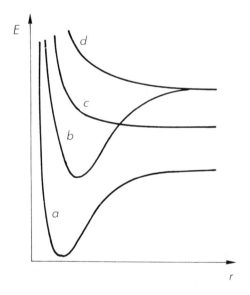

Fig. 15.4 — Morse curves corresponding to different electronic energy states.

### 15.1.5 Ionization
The Franck–Condon principle can similarly be applied to the process of ionization, which is also very fast, taking less than $10^{-16}$ sec. Thus **vertical ionization** can be defined in the same way as vertical transitions, and gives rise to molecular ions ($M^{+\cdot}$)

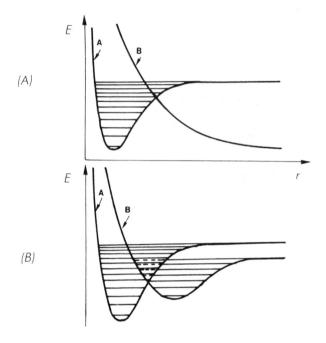

Fig. 15.5 — Intersection of different types of potential energy curves.

and fragment molecular ions ($^*M_1^+\cdot$ and $^*M_2^+\cdot$ in Fig. 15.6) in various states of excitation. Adiabatic ionization can also be analogously defined, as the conversion of the ground-state molecule into the ground-state molecular ion. Clearly the vertical ionization potential (the energy required for the ionization) will always be greater than the adiabatic ionization potential.

## 15.2  POLYATOMIC MOLECULES

### 15.2.1  Potential-energy surfaces

For polyatomic molecules potential-energy surfaces are used instead of potential-energy curves. Reactions taking place on a single potential-energy surface are called adiabatic; those taking place across different surfaces are called non-adiabatic. As with Morse curves, the surfaces for excited states can intersect the surface for the ground state. The probability of intersection increases with the number of surfaces, with the dimensions of the surfaces, and hence with the number of atoms in the molecule.

The position of a point on the surface defines the potential energy and the co-ordinates of the atoms. The path on the surface that corresponds to the changes in potential energy and position of the atoms of a molecule during a reaction is called a trajectory. Thus a reaction involving atomic displacement is visualized as a trajectory that begins in the domain of the reactants and ends in the domain of the products.

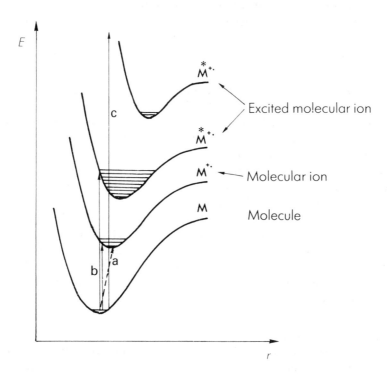

Fig. 15.6 — Adiabatic (a, broken line) and vertical (b and c, full lines) ionization.

The trajectory which follows the path of minimum potential energy is called the reaction co-ordinate.

For unimolecular and bimolecular reactions the potential energy surface will have a maximum between the reactant region and the product region, and this represents a potential energy barrier corresponding to the thermodynamics of the bond-breaking process. When two potential surfaces intersect, the resultant multi-potential surface will have a saddle, which represents both a minimum on the multipotential surface at the col between the two potential valleys on the reaction co-ordinate, and a maximum on the reaction profile.

The reaction rate is determined by the height of the saddle and the shape of the trajectory. The slope of the trajectory as it descends towards the products will determine their kinetic energy.

As an example, consider the bimolecular reaction $MN + X \rightarrow M + NX$. If the slope of the trajectory after the saddle is low, the separation of M from the transition-state complex $M \ldots N \ldots X$ will be comparatively slow, and the repulsion force responsible for displacement of M will be transmitted from X through N. Hence in calculation of the kinetic energy of M after its separation, the centre of mass of the transition-state complex should be taken into account. On the other hand, if the slope is steep, the displacement of M is rapid, and direct repulsion between M and X should be considered, rather than that transmitted through N (i.e. between NX and M). If the input energy is high enough, the trajectory no longer follows the surface, and both kinetic and potential energy are released during the reaction. Analogous considerations apply to a unimolecular reaction (fragmentation) such as $A \rightarrow B + C$.

### 15.2.2 Redistribution of excitation energy in the molecular ion
When they are formed, the molecular ions are in various energy states, but a redistribution of internal energy takes place after the ionization process, in two main ways. If the molecular ion formed is in the electronic ground state, coupling between the anharmonic oscillators and energy transfer will result in the ions populating the lowest vibrational levels of the electronic ground state. On the other hand, if the ions are formed in excited electronic states, the redistribution of energy will be determined by the intersections between the potential energy surfaces. As the transitions take place to lower and lower electronic states, the molecules will finally occupy a wide range of vibrational–rotational levels in the electronic ground state.

It is obvious that the number of surface intersections, and hence the rate of redistribution of energy, will increase with the number of atoms in the molecule. It is also possible that the energy transfer can be confined to particular groups of atoms instead of taking place in the whole molecule.

### 15.2.3 The rate constant
The ions produced in the ion source will have a range of energies (depending on the energy spread of the electron beam or other means of excitation) from $E$ to $E + \delta E$, and will therefore be in various vibrational energy states. The system can be described as a microcanonical distribution defined by a quantity called the density of states and denoted by $\rho(E)$. The quantity $\rho(E)\delta E$ gives the number of quantum states of the system possessing an energy in the range between $E$ and $E + \delta E$. Only ions in those states having a potential energy equal to the activation energy $E_0$ (defined as the potential energy at the saddle point) and a kinetic energy $\delta E_t$ can cross the potential barrier (tunnelling effects are ignored in the simple theory). Crossing the potential barrier can be regarded as a one-dimensional translational motion along the reaction co-ordinate. Since the kinetic energy of an activated ion crossing the potential barrier must arise from internal conversion of vibrational energy, the number of internal degrees of freedom of the activated ion will be decreased by one. The velocity of passage over the potential barrier can also be calculated from the kinetic energy. The rate constant can then be calculated from the density of states and the frequency of crossing the potential barrier, as follows.

The quantum mechanical expression for the kinetic energy $E_t$ of a particle in a one-dimensional box of length $a$ is

$$E_t = \frac{n^2 h^2}{8 \mu a^2}$$

where $n$ is the quantum number ($n = 1, 2 \ldots$), $h$ is Planck's constant, and $\mu$ is the mass of the particle (in this case the reduced mass of the particle that will be fragmented). It follows that the maximum number of translational states possible ($N$) will be given by

$$N = \frac{2a}{h}\sqrt{2\mu E_t}$$

and the density of translational states (the number per unit energy) will be

$$\rho(E_t) = dN/dE = \frac{a}{h}\sqrt{\frac{2\mu}{E_t}}$$

The energy of the activated complex will be divided between the activation energy ($E_0$), the translational energy, ($E_t$), and the energy in other degrees of freedom, ($E - E_0 - E_t$), so the number of states for the activated complex will be $\rho(E - E_0 - E_t)\rho(E_t)\delta E$, and the fraction of all the systems that will be in the activated complex configuration will be given by $\rho(E - E_0 - E_t)\rho(E_t)/\rho(E)$. The reaction rate will then be given by the product of this fraction and the frequency of crossing the saddle point. Since the frequency will be the velocity of the particle ($v$) divided by the distance traversed (in this case $a$) but only half the particles will be moving towards the saddle point, the frequency will be $v/2a$. The velocity is given by $v = \sqrt{2E_t/\mu}$, so $\rho(E_t)v/2a = h$. However, a system with total energy $E$ may dissociate with a kinetic energy $E_t$ having any value from zero to $E - E_0$, so the expression for the average rate constant will be

$$k = \int_0^{E - E_0} \frac{\rho(E - E_0 - E_t)}{h\rho(E)} dE_t$$

### 15.2.4  Competitive reactions
If several fragmentation paths are possible, each will have its own characteristic reaction co-ordinate and transition state, etc. The rate constant for each reaction will depend on the form of the potential surface and the height of the saddle point. Thus for the pair of fragmentations

$$A \underset{k_C}{\overset{k_B}{\lessgtr}} \begin{matrix} B \\ C \end{matrix}$$

the rate equation is

$$-d[A]/dt = (k_B + k_C)[A]$$

and at time $t$

$$[A] = [A]_0 e^{-(k_B + k_C)t}$$

and

$$d[B]/dt = k_B[A]_0 e^{-(k_B + k_C)t}$$

which on integration gives

$$[B] = \left(\frac{-k_B[A]_0}{k_B + k_C}\right) e^{-(k_B + k_C)t} + \text{constant}$$

If $[B] = 0$ at $t = 0$, then const $= k_B[A]_0/(k_B + k_C)$ and

$$[B] = \left(\frac{k_B[A]_0}{k_B + k_C}\right)[1 - e^{-(k_B + k_C)t}]$$

At any time $t$, $[A] + [B] + [C] = [A]_0$, so

$$[C] = \left(\frac{k_C[A]_0}{k_B + k_C}\right)[1 - e^{-(k_B + k_C)t}]$$

# 16

# Ion fragmentation mechanisms

## 16.1 FORMATION OF FRAGMENT IONS

The pattern of fragmentation of a molecular ion depends mainly on the chemical structure of the compound, and a number of empirical rules for predicting it have been established. Basically, the cleavages are of three main types: simple $\sigma$-bond cleavage, cleavage induced by the presence of a radical site, and cleavage induced by charge displacement. Obviously the pattern will also depend on the particular groups and bonds which can be the sites of these events. Some examples are given below, but represent only a small fraction of the possibilities.

### 16.1.1. Sigma-bond cleavages

$$(A\text{–}B\text{–}C\text{–}D) \xrightarrow{\;-e^-\;} (A\text{–}B\overset{+}{\cdot}C\text{–}D) \rightarrow AB^+ + \cdot CD$$

The site of ionization is the molecular orbital bonding atoms B and C. The loss of one of the electrons leads to cleavage of the B–C bond.

### 16.1.2 Fragmentation induced by a radical site ('r' type)

A radical site is generated during ionization, by the loss of an electron from a double bond or a hetero-atom. The resultant unpaired electron in the bond forms a new bond with an adjacent ($\alpha$) atom, and the fragmentation occurs between the $\alpha$ and $\beta$ atoms.

$$X\text{–}X'\text{–}\overset{\cdot+}{Y}\text{–}Z\text{–}Z' \xrightarrow[\alpha\text{-cleavage}]{\;r\;} X\cdot + X' = Y^+Z\text{–}Z'$$

### 16.1.3   Fragmentation by charge displacement–inductive cleavage ('*i*' type)

$$X\text{--}X'\text{--}\overset{+}{Y}\text{--}Z\text{--}' \xrightarrow{\quad i \quad} X\text{--}X'\text{--}Y + \overset{+}{Z}\text{--}Z'$$

Here the ionization potential of the $Z'\text{--}Z^+$ fragment is lower than that of the separated neutral $X\text{--}X'\text{--}Y$. The ionization potential (IP) of the $Z'Z^+$ ion fragment is lower than that of the neutral fragment: $IP(Z'Z^+) < IP(XX'Y)$.

### 16.2   FRAGMENTATION OF HYDROCARBONS

### 16.2.1   n-Alkanes
The ionization site is a C–C $\sigma$-bond. The mass spectrum contains peaks for ions separated by 14 mass units, corresponding to loss of $CH_2$. The intensities of the peaks are approximately related to the corresponding masses of the fragments. The molecular peak is not important when EI conditions are used.

### 16.2.2   Branched alkanes
The cleavage site will most probably be a $\sigma$-bond on the most highly substituted carbon atom, and involve the longest substituent (because this gives the greatest possibility of delocalization of the electron). The highest peak (for the example) corresponds to loss of $C_3H_7$. Figure 16.1 shows some typical mass spectra for branched alkanes.

$$C_2H_5\text{---}\underset{\underset{\displaystyle CH_3}{|}}{\overset{\overset{\displaystyle C_2H_5}{|}}{C}}\text{---}C_3H_7$$

### 16.2.3   Cycloalkanes
In EI these compounds exhibit a higher molecular peak than do the corresponding linear alkanes, and fewer fragment ions. Fragmentation of the ring is similar to that of the linear alkanes, resulting in peaks for ions differing by 14 mass units.

### 16.2.4   Polycyclic compounds
The pattern of cleavage is very complicated, and several fragmentation paths lead to a peak for the same $m/z$ value, corresponding to a variety of structures. For example, the mechanism given on page 140 can give rise to a peak at $m/z = 217$, but other fragmentations also yield the same peak. (Note that $\rightcurvearrow$ denotes shift of an electron pair, and $\rightharpoonup$ the shift of a single electron.)

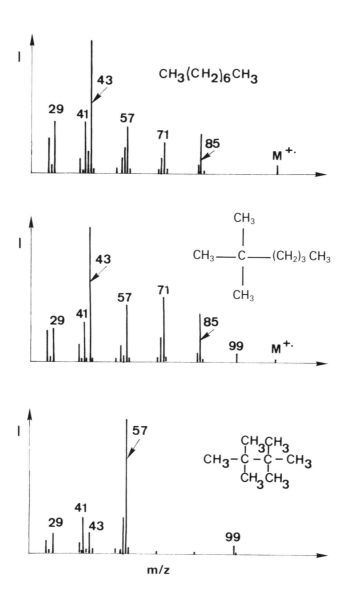

Fig. 16.1.

σ— clearage
−e⁻

r

r

m/z=217

### 16.2.5  Alkenes

Here the electron is lost from the double bond, and the presence of the radical site induces *r* type reactions. The alkene isomers have very similar mass spectra, and for this reason it is difficult to locate the positions of the double bonds by use of normal mass spectra.

−e⁻

r

stabilization

(+)

+ •CH₃

### 16.2.6  Cycloalkenes
(a) The simple case of a single double bond involves an *r* reaction followed by stabilization

(b) In the case of aromatic compounds, a characteristic peak is $m/z = 91$, corresponding to formation of $C_7H_7^+$.

### 16.2.7  Reactions of *i* type
This type of cleavage is less frequent than *r* cleavage (step II in the example). An *i* type reaction is step (I) of the following sequence. Its probability depends on the nature of R. For R = H it is about 5%, but 100% for $R = C_6H_5$.

## 16.3  FRAGMENTATION OF COMPOUNDS CONTAINING A HETERO-ATOM

### 16.3.1  Reactions of *r* type

The electron lost in the ionization comes preferentially from the lone pair of the hetero-atom. Four examples are shown below.

(1)

$$CH_3CH_2\overset{..}{O}C_2H_5 \xrightarrow[\text{ionization}]{-e^-} CH_3CH_2\underset{}{\overset{\overset{\bullet +}{}}{O}}C_2H_5$$

$$CH_2\!\!=\!\!\overset{+}{O}\!\!-\!\!C_2H_5 + {}^\bullet CH_3 \xleftarrow{\quad r \quad} CH_3\!\!-\!\!CH_2\!\!-\!\!\overset{\overset{\bullet +}{}}{O}\!\!-\!\!C_2H_5$$

stabilization

$$\overset{+}{H_2C}\!\!-\!\!O\!\!-\!\!C_2H_5$$

(2)

$$\underset{R_2}{\overset{R_1}{>}}\!\!C\!\!=\!\!O \quad (R_2\!>\!R_1)$$

$-e^-$ / ionization

$$\underset{R_2}{\overset{R_1}{>}}\!\!C\!\!=\!\!\overset{+}{O}{\overset{\bullet}{}} \longrightarrow \underset{R_2}{\overset{R_1}{>}}\!\!C\!\!=\!\!\overset{+}{O}{\overset{\bullet}{}}$$

*r*

$$R_1\!\!-\!\!\overset{+}{C}\!\!=\!\!O \xleftrightarrow{\text{stabilization}} R_1\!\!-\!\!C\!\!\equiv\!\!O^+ \quad + \quad R_2$$

Figure 16.2 shows the mass spectrum for $R_1 = C_2H_5$ and $R_2 = (CH_2)_2CH_3$, the corresponding ions having $m/z = M - 29$ and $M - 43$.

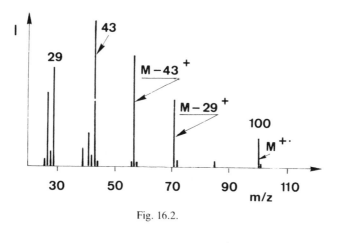

Fig. 16.2.

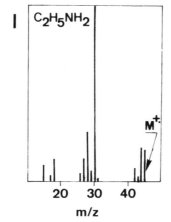

Fig. 16.3.

(3)

$$R\!-\!CH_2\!-\!NH_2 \xrightarrow[-e^-]{\text{ionization}} R\!-\!CH_2\!-\!\overset{\bullet\,+}{NH_2} \longrightarrow R\!-\!CH_2\!-\!\overset{\bullet\,+}{NH_2}$$

$$\downarrow r$$

$$CH_2\!=\!\overset{+}{N}H_2 + R$$

$$m/z = 30$$

Figure 16.3 shows the mass spectrum for ethylamine; the fragmentation to yield $CH_2 = \overset{+}{N}H_2$ is clearly evident.

(4)

$$C_3H_7\!-\!\underset{\underset{C_2H_5}{|}}{\overset{\overset{CH_3}{|}}{C}}\!-\!\overset{\bullet\,+}{OH} \longrightarrow C_2H_5C(CH_3)\!=\!\overset{+}{OH} + \overset{\bullet}{C_3H_7}$$

$$m/z = 73$$

The loss of $C_2H_5$ instead of $C_3H_7$ gives ions with $m/z = 87$, as shown in Fig. 16.4.

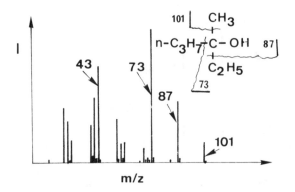

Fig. 16.4.

### 16.3.2  Fragmentation determined by the charge (*i* type)

(1)         $CH_2\overset{+}{=\!\!\!=}O\!-\!\!C_2H_5$         $\xrightarrow{\;\;i\;\;}$         $C_2H_5{}^+ + CH_2O$

Note that (EE)→(EE)+(EE) where (EE) means an even-electron species,
IP(C$_2$H$_5$ < IP(CH$_2$O).

(2)         $R\!-\!\underset{\overset{\|}{\underset{+\bullet}{O}}}{C}\!-\!NHR'$         $\xrightarrow{\;\;r\;\;}$         $RC\overset{+}{\equiv\!\!\!=}O + \dot{N}HR'$

$$\downarrow$$

$$\overset{+}{R} + CO$$

An *r* reaction is followed by an *i* reaction, in the case of peptides.

Note that (OE) $\overset{r}{\rightarrow}$ (EE) $\overset{i}{\rightarrow}$ (EE), where (OE) means an odd-electron species,
(EE) an even-electron species, and IP(R) < IP(CO).

(3)

*m/z*=66

## 16.4   REARRANGEMENTS

### 16.4.1   Transfer of a hydrogen atom, determined by a radical site
(1)  Transfer to a radical site (*tH*)

ionization
$-e^-$

*tH*

*tH*

$m/z=218$

$+ \; CH_2 = CHCH_2R$

*r*

(2)  Transfer to a hetero-atom with a double bond (the McLafferty rearrangement)

The transfer takes place through a six-membered ring transition state, with β-cleavage.

$CH_3 \!\!-\!\!(CH_2)\!\!-\!\!\overset{\overset{\bullet+}{O}}{\underset{\|}{C}}\!\!-\!\!CH_3$

*tH*

$+ \; C_2H_4$

stabilization

*r*

**(3)** Transfer to a saturated hetero-atom.

### 16.4.2 Transfer of a hydrogen atom, determined by the charge
This generally occurs through a four-membered transition state.

### 16.4.3 Displacement reactions (*d*)
These involve the cleavage of one bond and the formation of a new bond between
two distant atoms. They are initiated by (a) a radical site or (b) a charge:

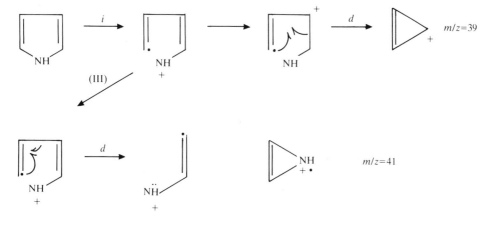

(2) This example follows the evolution of the ion formed by transfer of a hydrogen atom to a saturated nitrogen hetero-atom (see Section 16.4.1, example (3), sequence I)

In sequence II in the same example a displacement reaction also occurs.

(3) In this example *r* fragmentation is followed by transfer of a hydrogen atom and a displacement reaction.

$m/z = 133$

### 16.4.4  Elimination reactions
In elimination reactions a neutral situated between two atoms is eliminated from the ion. Two bonds are cleaved.

## 16.5   INTERPRETATION OF MASS SPECTRA [1–4]

As can be seen above, fragmentation patterns can be both variable and complicated for the same species, and the mechanisms proposed are not always predictable from the structure of the parent molecule. In the case of an 'unknown' compound, the structure has to be deduced from the mass spectrum, and there are well established guidelines for doing this. The obvious first step is to identify the parent molecular ion $M^{+\cdot}$, followed by identification of the main peaks. If the spectrometer used does not supply a print-out in the form of a bar-graph (such as Fig. 1.1), the intensities must be measured and normalized. The abundance and distribution of the ions in the spectrum will give information about the size and stability of the compound. Characteristic peaks will be associated with fragments such as $CH_2NH_2^+$ ($m/z = 30$) from amines, and $C_6H_5^+$ ($m/z = 77$) from phenyl compounds, and so on. Metastable peaks should then be noted and the decomposition identified by specific methods. Collisionally activated decomposition of the ions corresponding to the main peaks should be performed. Neutral fragments which are lost can be correlated with high-mass peaks such as $(M-1)^+$, $(M-15)^+$, $(M-18)^+$, $(M-20)^+$, representing loss of H, $CH_3$, $H_2O$ and HF, respectively, by means of correlation tables. Likely single-bond cleavages can be postulated on the basis of the enthalpies of formation of the fragments produced.

These and other rules, together with other spectroscopic information (NMR, UV, IR, etc.) are all brought into play in identification of the compound. The reader is referred to the specialist works on this topic for further information and practice in interpretation.

## REFERENCES

[1] F. W. McLafferty, *Interpretation of Mass Spectra*, 3rd Ed., University Science Books, Mill Valley, CA, 1980.
[2] M. E. Rose and R. A. W. Johnstone, *Mass Spectrometry for Chemists and Biochemists*, Cambridge University Press, Cambridge, 1982.
[3] Q. N. Porter, *Mass Spectrometry of Heterocyclic Compounds*, 2nd Ed., Wiley, New York, 1985.
[4] M. C. Hamming and N. G. Foster, *Interpretation of Mass Spectra of Organic Compounds*, Academic Press, New York, 1972.

# 17

## Isotopic analysis

### 17.1 ISOTOPIC COMPOSITION OF THE ELEMENTS

Chemical elements can have several isotopes, but sometimes only one or a few of them will be stable, and for some elements none occur in nature. Thus naturally occurring oxygen contains isotopes with mass numbers 16, 17 and 18, and magnesium also has three stable isotopes, with mass numbers 24, 25 and 26. The isotopic compositions of these two elements are as follows.

| Element | Mass number | Relative abundance, % |
|---------|-------------|----------------------|
| Oxygen | 16 | 99.76 |
| | 17 | 0.04 |
| | 18 | 0.20 |
| Magnesium | 24 | 78.99 |
| | 25 | 10.00 |
| | 26 | 11.01 |

In contrast, elements such as fluorine, phosphorus, and manganese have only one stable isotope.

| Element | Mass number | Exact atomic mass |
|---------|-------------|-------------------|
| fluorine | 19 | 18.9984 |
| Phosphorus | 31 | 30.9738 |
| Manganese | 55 | 54.938 |

Since the isotopes of an element have different mass numbers, so will the ions derived from them. For example, if krypton (which has monatomic molecules) is introduced into the ion source, the peaks in the mass spectrum will be located at the following $m/z$ values and relative intensities:

| $m/z$ | 78 | 80 | 82 | 83 | 84 | 86 |
|---|---|---|---|---|---|---|
| Relative intensity, % | 0.36 | 2.27 | 11.56 | 11.55 | 56.90 | 17.37 |

If an element such as chlorine, which has two stable isotopes and forms a diatomic molecule, is ionized to produce the molecular ion, the peaks will correspond to the various combinations of the isotopes. The isotopes are $^{35}Cl$ (abundance 24.5%) which can combine to yield

| | |
|---|---|
| $^{35}Cl^{35}Cl^+$ | $m/z = 70$ |
| $^{35}Cl^{37}Cl^+$ | $m/z = 72$ |
| $^{37}Cl^{37}Cl^+$ | $m/z = 74$ |

For an ion containing atoms of two elements, all the possible combinations of the isotopes must be taken into account, in accordance with their relative abundances. For carbon dioxide, for example, the isotopes to be considered are $^{12}C$ (92.892%), $^{13}C$ (1.108%), $^{16}O$ (99.76%), $^{17}O$ (0.04%) and $^{18}O$ (0.20%), and in the mass spectrum of $CO_2^+$ peaks will appear at $m/z$ values corresponding to the following nominal masses between 44 and 49:

| | |
|---|---|
| $^{12}C^{16}O^{16}O$ | $m = 44$ |
| $^{12}C^{16}O^{17}O$ or $^{13}C^{16}O^{16}O$ | $m = 45$ |
| $^{12}C^{16}O^{18}O$ or $^{12}C^{17}O^{17}O$ or $^{13}C^{16}O^{17}O$ | $m = 46$ |
| $^{13}C^{18}O^{18}O$ | $m = 49$ |

Figure 17.1 shows the mass spectrum for a compound which yields molecular peaks

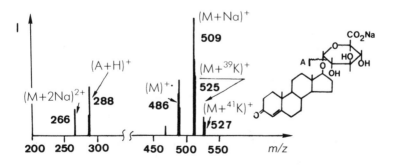

Fig. 17.1 — Mass spectrum showing the effect of the isotopes of potassium on cationization. (Reproduced by permission, from H. R. Schulten and D. E. Games, *Biomed. Mass Spectrom.*, 1974, **1**, 120. Copyright 1974, John Wiley & Sons, Inc., New York.)

cationized with potassium; the peaks for the ions formed with the two isotopes of potassium are clearly seen, and have intensities in the same ratio as the abundances of the two isotopes.

## 17.2   CALCULATION OF SPECTRAL DISTRIBUTION OF ISOTOPIC COMBINATIONS, RELATIVE MASSES AND INTENSITIES

Let us take carbon as an example of an element with more than one stable isotope. The possible isotopic combinations for two carbon atoms in a compound are

$$^{12}C^{12}C \quad ^{12}C^{13}C \quad ^{13}C^{12}C \quad ^{13}C^{13}C$$

giving four possibilities:

$$(^{12}C^{12}C) + 2(^{12}C^{13}C) + (^{13}C^{13}C) \tag{17.1}$$

Likewise, for three carbon atoms, since the additional atom may be either $^{12}C$ or $^{13}C$, there are the possibilities:

$$^{12}C^{12}C^{12}C \quad ^{12}C^{12}C^{13}C \quad ^{12}C^{13}C^{12}C \quad ^{13}C^{12}C^{12}C$$
$$^{12}C^{13}C^{13}C \quad ^{13}C^{12}C^{13}C \quad ^{13}C^{13}C^{12}C \quad ^{13}C^{13}C^{13}C$$

a total of eight:

$$(^{12}C_3) + 3(^{12}C_2{}^{13}C) + 3(^{12}C^{13}C_2) + (^{13}C_3) \tag{17.2}$$

If we denote $^{12}C$ by $a$ and $^{13}C$ by $b$ and write the binomial expansion $(a + b)^n$

$$a^2 + 2ab + b^2 \text{ for } n = 2$$
$$a^3 + 3a^2b + 3ab^2 + b^3 \text{ for } n = 3$$

and equate these expressions with (17.1) and (17.2), we see that the coefficients of the terms in the binomial correspond to the numbers of a given combination of $^{12}C$ and $^{13}C$ atoms, the relative numbers of each isotope in a combination being given by the power to which $a$ or $b$ is raised. The relative abundances (and hence the relative peak intensities for the different isotopic combinations) can be calculated in similar fashion from the abundances of the two isotopes ($^{12}C$ 98.89%, $^{13}C$ 1.11%).

  If we take the example of a molecule containing three chlorine atoms, we obtain the same distribution of the combinations as for the carbon example above, and if we denote $^{35}Cl$ by $a$ and $^{37}Cl$ by $b$, and the mass of the molecule containing three $^{35}Cl$ atoms by M, we can construct a table showing the possible peaks appearing in the mass spectrum, and their relative intensities. This example is simplified by the fact that the abundance ratio $^{35}Cl/^{37}Cl = 75.5/24.5$ is approximately equal to 3.

| Binomial term | Isotope combination | Mass | Relative intensity |
|---|---|---|---|
| $a^3$ | $^{35}Cl_3$ | M | $3^3 = 27$ |
| $3a^2b$ | $^{35}Cl_2{}^{37}Cl$ | M + 2 | $3 \times 3^2 \times 1 = 27$ |
| $3ab^2$ | $^{35}Cl^{37}Cl_2$ | M + 4 | $3 \times 3 \times 9 = 9$ |
| $b^3$ | $^{37}Cl_3$ | M + 6 | $1^3 = 1$ |

An analogous argument is used for substitution by more than one polyisotopic element, e.g. two chlorine atoms ($a = {}^{35}Cl$ and $b = {}^{37}Cl$) and one of bromine ($c = {}^{79}Br$ and $d = {}^{81}Br$, abundances 50.54 and 49.46% respectively). The algebraic expression in this case is

$$(a + b)^2(c + d) = a^2c + 2abc + b^2c + a^2d + 2abd + b^2d$$

If M is the mass of the $^{35}Cl_2\,{}^{79}Br$ combination, we obtain:

| Algebraic term | Isotope combination | Mass | Relative intensity | |
|---|---|---|---|---|
| $a^2c$ | $^{35}Cl\,_2{}^{79}Br$ | M | $3^2 \times 1 = 9$ | |
| $2abc$ | $^{35}Cl^{37}Cl^{79}Br$ | M + 2 | $2 \times 3 \times 1 \times 1 = 6$ | } 15 |
| $a^2d$ | $^{35}Cl_2{}^{81}Br$ | M + 2 | $3^2 \times 1 = 9$ | |
| $b^2c$ | $^{37}Cl_2{}^{79}Br$ | M + 4 | $1^2 \times 1 = 1$ | } 7 |
| $2abd$ | $^{35}Cl^{37}Cl^{81}Br$ | M + 4 | $2 \times 3 \times 1 \times 1 = 6$ | |
| $b^2d$ | $^{37}Cl_2{}^{81}Br$ | M + 6 | $1^2 \times 1 = 1$ | |

The appearance of the mass spectrum is shown in Fig. 17.2.

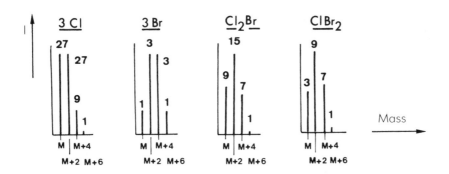

Fig. 17.2 — Relative intensities of peaks corresponding to different isotopic combinations.

## 17.3  MOLECULAR REGIONS AND ISOTOPIC DISTRIBUTION

From the details given above, it is clear that the treatment can be extended to more complicated molecules, but the calculations become increasingly complex. For organic molecules, however, the mass numbers of the isotopes of a given component element generally cover only a small numerical range, and for many of the constituent elements one isotope will be predominant (e.g. $^{12}C$, 98.89%, $^{1}H$ 99.98%, $^{14}N$ 99.63%, $^{16}0.99.76\%$) and one isotopic combination will also be predominant. There will thus be a major peak accompanied by a number of minor peaks within a narrow mass range. Some of these peaks will arise from different combinations of isotopes that happen to give the same mass number, as shown above for the case of a compound containing two chlorine atoms and one bromine atom (M + 2 and M + 4). For a compound $C_wH_xN_yO_z$ the ratio of the intensities for the isotopic combinations can be calculated from formulae [1] such as

$$\frac{P_{M+1}}{P_M} = w\left(\frac{c}{100-c}\right) + x\left(\frac{h}{100-h}\right) + y\left(\frac{n}{100-n}\right) + z\left(\frac{o}{100-o-o'}\right)$$

where $c$, $h$, $n$, $o$ and $o'$ are the isotopic abundances of $^{13}C$, $^{2}H$, $^{15}N$, $^{17}O$ and $^{18}O$. $P_{M+1}$ is the intensity for the peak corresponding to an ion containing only the most abundant isotope ($^{12}C$ etc.). Corresponding formulae can be devised for ions of mass M + 2 and so on, but rapidly become too complicated for easy calculation. However, high-resolution mass spectrometers allow the separation of multiplets and the experimental determination of the relative intensities of the peaks corresponding to different isotopic combinations, and computer programs have been written for calculation of the exact masses of these combinations [2].

**REFERENCES**

[1] J. H. Beynon, *Mass Spectrometry and its Applications to Organic Chemistry*, Elsevier, Amsterdam, 1960, p. 296.
[2] D. D. Tunnicliff, P. A. Wadsworth and D. O. Schissler, *Anal. Chem.*, 1965, **37**, 543, and references therein.

# 18

# Recent developments

## 18.1  DESORPTION AND IONIZATION METHODS

Research into high molecular-weight molecules is one of the aims of modern spectrometry. The difficulties encountered in this field stem partly from the unstable nature of many of these molecules and partly from the characteristics of mass analysers, the performance of which decreases markedly as the mass to be measured increases.

The instability of the molecule causes it to decompose either before ionization, as it passes into the gas phase, or after ionization, as a result of the excitation energy acquired. This means that the peak of the parent molecule, which is fundamental for a successful total analysis since it gives the molecular weight, is absent from the mass spectrum.

However, special ionization–desorption methods are now available which allow survival of the intact parent molecule, so that the mass spectrum shows the peak for the molecular ion as well as the peaks for the fragment ions (see Chapter 13). These methods all give 'soft' ionization, with minimal fragmentation. Figure 18.1 shows the molecular peak region in the mass spectrum (obtained by FAB) of a glycopeptide (fibrinogen, $C_{77}H_{126}N_7O_{56}$).

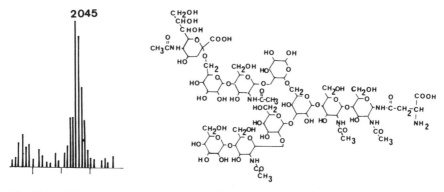

Fig. 18.1 — Mass spectrum of a glycopeptide (fibrinogen). (Reprinted by permission, from C. Fenselau, R. J. Cotter, D. Heller and J. Yergey, *J. Chromatog.*, 1983, **271**, 3. Copyright 1983, Elsevier, Amsterdam.)

Figure 18.2 shows two mass spectra for chlorophyll (M = 892), a molecule which is thermally unstable. The spectra were obtained with electron impact ionization of chlorophyll on a gold sample holder, and show the evolution of the spectrum in successive scans (cf. Section 7.3). The molecular peak $M^+$ is present in the first scan, but disappears in later scans, which give only the peaks for the fragment ions.

The study of amino-acid sequences in peptides is one in which mass spectrometry is of great importance. The first noteworthy results were obtained by field desorption (FD, Section 2.8) and SIMS (Section 2.10), and with a gold sample holder. Most of the results are obtained by using the FAB method for ionization.

An important field has been opened up by the development of multiphoton ionization, which allows control of the degree of fragmentation by variation of the number and energy of the photons transferred to the molecule. Figure 18.3 shows the fragmentation patterns obtains for benzene by irradiating it with light of different wavelengths.

## 18.2 PROGRESS IN DEVELOPMENT OF MASS ANALYSERS

The most recent developments in the construction of mass analysers include (a) the use of laminated magnets (e.g. with laminae 0.3 mm thick in VG analysers) for the elimination of hysteresis, (b) increase in scan rate (to 0.05–0.1 sec per decade), (c) design of high-field magnets and use of an inhomogeneous magnetic field to extend the accessible mass range, (d) use of hexapole lenses at the entrance and exit of the magnetic sector, to eliminate or greatly reduce stray fields, allowing the use of wider slits and hence improving the sensitivity, especially for high resolution, (e) elimination of mechanical slits for regulating resolution. These improvements have increased the width of the mass region that can be scanned, the sensitivity, and the precision in mass determinations. As an example, Fig. 18.4 shows the molecular peak region in the mass spectrum of bovine insulin ($C_{254}H_{377}N_{65}O_{75}S_6$, molecular weight 5730.6) obtained with a VG ZAB mass spectrometer. The spectrum of human proinsulin ($C_{410}H_{638}N_{114}O_{127}S_6$, molecular weight 9388.6) has also been obtained, with a ZAB-HF spectrometer and a FAB source.

Progress has also been made in the quality of Fourier transform cyclotron resonance spectrometers and time-of-flight analysers. Ion reflectors have been added to the latter to increase the resolution and allow the separation of higher masses. Time-of-flight analysers are very useful when californium or laser ion-sources are used. Figure 18.5 shows part of the mass spectrum of a dimer (molecular weight 12550), and Fig. 18.6 shows the corresponding region of the spectrum of the monomer (molecular weight 6275) together with the structure of the molecule. With Fourier transform cyclotron resonance spectrometers and a laser desorption source a range of up to $m/z = 7000$ has been attained.

## 18.3 ANALYSIS OF MOLECULAR STRUCTURE

### 18.3.1 Fragment ions

The presence of fragment ion peaks in the mass spectrum is important because the fragments are produced by rupture of the weakest bonds, and thus yield information

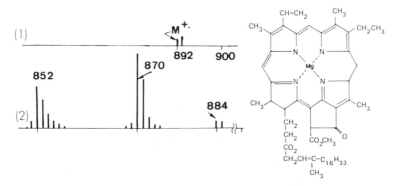

Fig. 18.2 — Mass spectrum of chlorophyll.

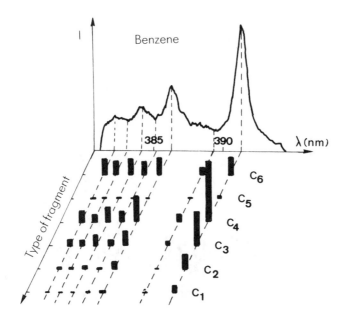

Fig. 18.3 — Mass spectra of benzene excited with light of different wavelengths. (Reproduced by permission, from L. Zandee and R. B. Bernstein, *J. Chem. Phys.*, 1979, **71**, 1359. Copyright 1979, American Physical Society, New York.)

about the structure of the parent molecule. Modern methods of investigating molecular structure make use of the information thus provided by the spontaneous decomposition of unstable ions after the ion source. When such decompositions do not normally occur, they can be induced by introduction of a collision gas into the ion path. Study of the fragments thus obtained and those from spontaneous

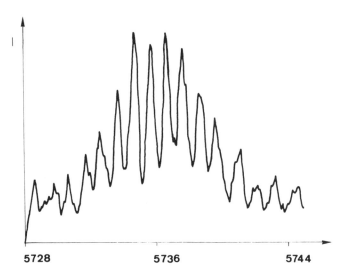

Fig. 18.4 — Molecular region in the spectrum of bovine insulin. (After J. C. Bill, B. N. Green and P. W. Brooks, 30th ASMS Conference, Hawai, 1982.)

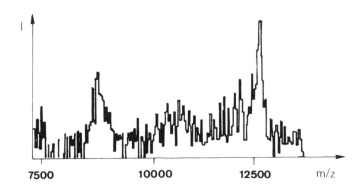

Fig. 18.5 — Mass spectrum of a dimer with molecular weight 12550. (Adapted, with permission, from C. J. McNeal and R. D. Macfarlane, *J. Am. Chem. Soc.*, 1981, **103**, 1609. Copyright 1981, American Chemical Society, Washington D.C.)

decomposition allows the structures of the precursor species to be deduced (see Chapters 10 and 16).

Broadly speaking, there are two main approaches, based on use of either one two-sector spectrometer having inverse or normal geometry, with or without a

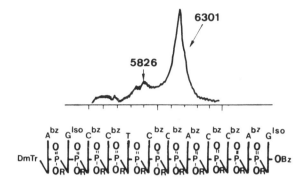

Fig. 18.6 — Mass spectrum of the monomer (of the dimer in Fig. 18.5), molecular weight 6275. (Adapted, with permission, from C. J. McNeal and R. D. Macfarlane, *J. Am. Chem. Soc.*, 1981, **103**, 1609. Copyright 1981, American Chemical Society, Washington D.C.)

reagent gas in the collision chamber, as discussed in Chapter 10, or two spectrometers in series (the so-called tandem or MS–MS method), again with or without a gas in the collision chamber.

### 18.3.2    The MS–MS method
The tandem system can be constructed with either quadrupole or magnetic analysers. The advantage of the quadrupole system is that it can deal with low-energy ions. Magnetic analysers, however, give much higher resolution.

Figure 18.7 shows the scheme for a quadrupole system in which the collision

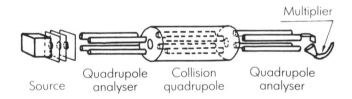

Fig. 18.7 — A quadrupole tandem mass spectrometer. [Reproduced by permission, from R. A. Yost and C. G. Enke, in *Tandem Mass Spectrometry*, F. W. McLafferty (ed.), p. 176. Copyright 1983, John Wiley & Sons, Inc., New York.]

chamber is itself a quadrupole, and Fig. 18.8 shows the design of a high-resolution tandem magnetic analyser instrument. There are also commercial spectrometers equipped with four analyser sectors, arranged in the order magnetic–electrostatic–electrostatic–magnetic, with a collision chamber in a field-free region.

Other commercial systems (for example the ZAB series E) can be used in the following configurations:

magnetic sector–electrostatic sector–quadrupole–quadrupole

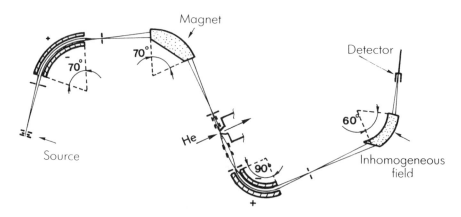

Fig. 18.8 — A high-resolution mass spectrometer. (Adapted by permission, from P. J. Todd, D. C. McGilvery, M. A. Baldwin and F. W. McLafferty, in *Tandem Mass Spectrometry*, F. W. McLafferty (ed.), p. 272. Copyright 1983, John Wiley & Sons, Inc., New York.)

magnetic sector–electrostatic sector–electrostic sector–magnetic sector

The maximum resolution is 125000 for the primary beam, and the mass range accessible is up to $m/z = 15000$.

### 18.3.3 Quality critical for a mass spectrometry system
The quality of a system for molecular analysis by mass spectrometry is currently judged by the following criteria:

the mass range;
the number of ionization methods available and the ease of switching between them;
the sample size, and whether the sample can be a raw material or needs to be separated, purified or derivatized;
the possibility of obtaining structural information, tracing the 'family tree' of the fragments, and analysing the fragmentation pattern;
the possibility of analysing mixtures;
availability of computerized data-acquisition and interpretation (by suitable software, spectral-library searches, etc., and the number and speed of the programs available);
final presentation of the results;
reproducibility of the results.

### 18.4   TOWARDS A NEW ANALYTICAL CHEMISTRY [1]

The immediate result of studying metastable decompositions or using MS–MS is the possibility of analysing very complex mixtures without prior chemical separation.

Generally, when the aim of the work is not to identify all the compounds in a mixture, but to look for only one of them, mass spectrometry eliminates or at least largely reduces the extraction or purification of the compound in question. A new analytical chemistry has thus developed, the possibilities of which have not yet been fully exploited.

Below are presented a number of results showing the progress that has been made in the fields of high molecular-weight determination, structural analysis, and trace detection.

## 18.5  EXPERIMENTAL RESULTS

Ionization of a mixture of steroids isolated from *Eunicella stricta* gives peaks appearing at $m/z$ = 428, 426, 414, 412, 400, 398, 386 and 384 (Fig. 18.9). Analysis of

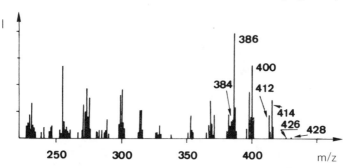

Fig. 18.9 — Mass spectrum of mixture of sterols isolated from *Eunicella stricta*. (Reproduced by permission, from A. Maquestio, Y. van Haverbeke, R. Flammang, H. Mispreuve, M. Kaisin, J. C. Braekman, D. Daloze and B. Tursch, *Steroids*, 1978, **31**, 31. Copyright 1978, Holden-Day, Inc., San Francisco.)

the metastable decompositions of some of these species and comparison with the parent peaks shows that the latter are in fact the molecular peaks for the different steroids in the mixture. Figure 18.10 shows the spectra from the spontaneous metastable decomposition of three of these ion species.

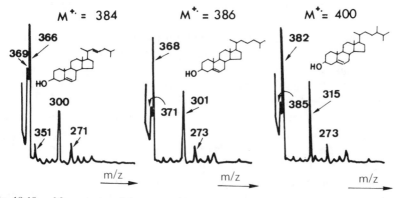

Fig. 18.10 — Mass spectra of the metastables from three of the products shown in Fig. 18.9. (Reproduced by permission, from A. Maquestio, Y. van Haverbeke, R. Flammang, H. Mispreuve, M. Kaisin, J. C. Braekman, D. Daloze and B. Tursch, *Steroids*, 1978, **31**, 31. Copyright 1978, Holden-Day, Inc., San Francisco.)

The same steroids are found in other marine invertebrates such as *Clavularia inflata*, *Alcyonum digitatum*, *Spongilla lacustris*, *Lemnalia digitata* and *Capnella imbricata* [2]. The amount of sample needed for these analyses was about 1 mg.

In certain cases the mixture consists almost entirely of one steroid. For example, for the sponge *Strongylophora durissima dendy* from the Indo-Pacific Ocean, 90% of the steroid mixture isolated was a new type of steroid with an unsaturated side-chain [3].

Another example is the discovery and identification of 16-keto-oestradiol and 16-hydroxyoestrone in fungal growths. For *in vitro* cultures of *Leptosphaeria typhae* [4], *Gnomonia leptostyla* [5] and *Candida albicans* [6] it has been observed that the ratios between steroids in them are very sensitive to the environment, especially radiation, temperature etc.

Mass spectrometry has also been used for identification of alkaloids. Thus chemical ionization of a raw extract from the cactus *Dolicothele longimamma* gave a peak at $m/z = 166$, and a study of the metastable decompositions of this ion indicated it had the structure shown as (a) in Fig. 18.11. A new alkaloid found by study of metastables occurs in *D. uberiformis*, and has the structure (molecular weight 193) shown as (b) in Fig. 18.11. The limit of detection was 1–10 ng.

Fig. 18.11 — Structure of compounds giving rise to peaks at (a) $m/z = 166$ for an extract from *D. longimamma* and (b) $m/z = 193$ for an alkaloid from *D. uberformis* [7].

Several studies have been made of mushrooms. Figure 18.12 shows a comparison of the MIKES spectrum of muscarine (A) with that of an untreated mushroom sample (B).

Another interesting area of study is the family of pyrrolizidine alkaloids (PAs), which have very varied molecular structure [8]. It is known that the presence of a 1,2 double bond makes this family hepato-toxic, but nothing else is known about the structure–activity relationships. Figure 18.13 shows the structures of some molecules of this type. Molecule A (molecular weight 337) is non-toxic, B (m.w. 325) is toxic, and C (m.w. 351) is the most toxic of all. These compounds are present in various plants, and pass through the food chain into animals and humans. These types of compound are present in certain kinds of tea and honey, for example, at 1–4 µg/g levels. Their effect on the human or animal organism is cumulative, and death can occur at any time once the critical dose has been reached.

To this picture must be added the fact that the lack of any apparent symptoms increases the risk of poisoning of humans by PAs in food. A large number of animals die in the United States from PA poisoning; hence the need to identify and find the characteristics of this type of molecule. In the United States, three types of plant are

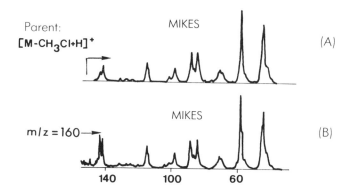

Fig. 18.12 — Mass spectrum of (A) muscarine, (B) MS–MS spectrum for an untreated mushroom sample. (Reproduced by permission, from S. E. Unger, A Vince, R. G. Cooks, R. Chrisman and L. D. Rothman, *Anal. Chem.*, 1981, **53**, 976. Copyright 1981, American Chemical Society, Washington D.C.).

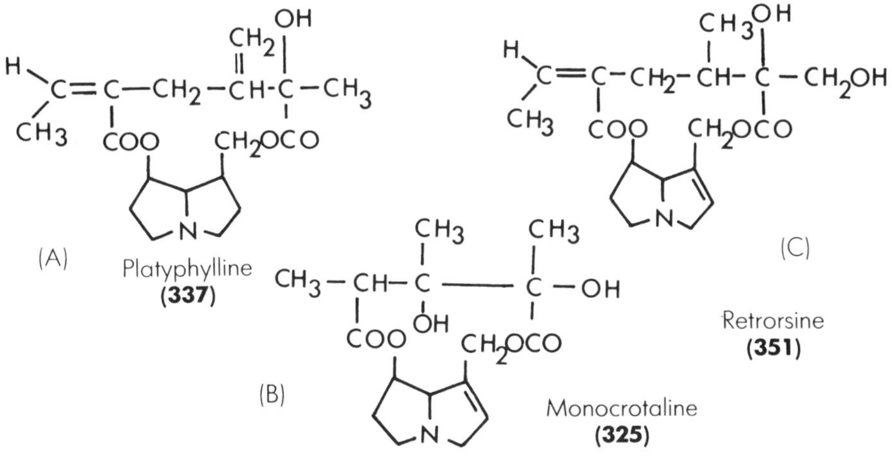

Fig. 18.13 — Three pyrrolizidine alkaloids identfied by mass spectrometry.

especially dangerous in this connection: *Senecio douglasii, Senecio jacobaea* and *Senecio riddellii*, in which the alkaloids may amount to 10% or more of the total weight of the plant.

The analyses were done with the dried plant or a methanol extract, by chemical ionization. The compounds were identified by using combined spectra of $B/E$ and $B^2/E$ type (Sections 10.3.3 and 10.3.4) for examination of the metastables. In the

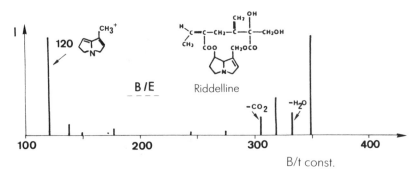

Fig. 18.14 — Mass spectrum of riddelline. (Reproduced by permission, from W. F. Haddon and R. J. Molyneux, in *Tandem Mass Spectrometry*, W. F. McLafferty (ed.), p. 459. Copyright 1983, John Wiley & Sons, Inc., New York.)

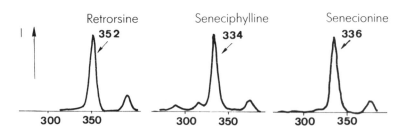

Fig. 18.15 — $B^2/E$ peaks for retrorsine, seneciphylline and senecionine. (Reproduced by permission, from W. F. Haddon and R. J. Molyneux, in *Tandem Mass Spectrometry*, W. F. McLafferty (ed.), p. 459. Copyright 1983, John Wiley & Sons, Inc., New York.)

Fig. 18.16 — $B^2/E$ peaks for the quasi-molecular ion of M + H = 350 and for $m/z$ = 120 for certain *Senicio* species. (Reproduced by permission, from W. F. Haddon and R. J. Molyneux, in *Tandem Mass Spectrometry*, W. F. McLafferty (ed.), p. 459. Copyright 1983, John Wiley & Sons, Inc., New York.)

spectrum shown in Fig. 18.14, the characteristic fragment peak is at $m/z = 120$. To characterize the mixture (from the methanol extract or the raw material), all the precursors which could generate an ion with $m/z = 120$ were sought by the $B^2/E$ method. Several such precursors exist (as shown in Fig. 18.15), corresponding to quasi-molecular ions with $M + H = 352, 334$ and $336$, which are attributable to PAs.

These $B^2/E$ analyses confirmed the presence of PAs in six species of plants. *S. riddellii*, *S. multicapitus* and *S. blanchmanae* were shown to contain a compound, which yielded a quasi-molecular ion with $M + H = 350$ (Fig. 18.16).

Another example is that a 1-mg sample of commercial tea was shown to contain caffeine, by introducing it directly into a chemical ionization source. The decomposition spectrum induced by collision (for ions with $m/z = 195$) was compared with that obtained from pure protonated caffeine under the same conditions.

Similarly useful results have been obtained in analysis for nicotine ($m/z = 163$) in tobacco, or for various characteristic compounds in *Conium maculatum*, *Papaver somniferum L.*, *Erythoxylum coca Lam.*, walnuts, Indian ink etc.

For the analysis of plants a simple method of sample preparation is freezing in liquid nitrogen, followed by crushing and pulverization.

MS–MS analysis has been applied to human serum, urine samples and various other body fluids in pharmacokinetic research. The detection limits are as low as 1–10 ng/ml.

An atmospheric pressure source has been used to detect contamination in food, e.g. tetrachlorodibenzo-*p*-dioxin in fish, mycotoxins in cereals and meat, pesticides and herbicides in vegetables, and chlorinated phenols in chickens.

An interesting application is the detection of contraband drugs or explosives. This can be done in two ways. The first is head-space analysis of the air in the container that is suspect, and the other is to subject the suspected object or person to a current of gas, which is then analysed. Letters, baggage, containers and people or animals can be examined in this way. Volatile compounds, such as residual acetic acid in processed heroin, methyl benzoate in cocaine, and nitroglycerine, can be detected immediately, without preconcentration.

Less volatile compounds, such as heroin itself, or TNT, require preconcentration, which can be achieved by passing the gas current over a wire previously coated with OV-101 as a stationary phase. The accumulated analyte can then be desorbed by heating the wire electrically.

Fruit or meat smuggled in baggage can be detected by means of the terpenes, terpinols, lactones, esters, alcohols, amines, sulphides, etc. that are present in them. The detection limits for a number of classes of compound are listed in Table 18.1.

The methods just described have been used to find and identify the major constituents of medicinal plants and for revealing the presence or absence of a particular compound, and to obtain the mass spectrum of individual biological particles (e.g. bacteria). For the latter purpose, the particles are introduced into the ion source as an aerosol made by nebulization of a suspension in methanol. Evaporation of the solvent and ionization of the particle take place on impact with a heated rhenium filament (Fig. 18.17). Examples are the examination of *B. cereus*, *B. subtilis* and *Pseudomonas putida*. Exhaled gases can also be analysed by atmospheric pressure mass spectrometry (TAGA, Section 13.2.3). A main aim is the determi-

**Table 18.1** — Limits of detection for some typical compounds in the gas phase (information by courtesy of SCIEX Inc., Thornhill, Ontario)

| Substances | Limit of detection |
|---|---|
| Benzaldehyde | $10 \, nl/m^3$ |
| Toluene | $5 \, nl/m^3$ |
| Dichlorobenzene | $1 \, nl/m^3$ |
| Diethyl ether | $50 \, nl/m^3$ |
| Anthracene | $50 \, nl/m^3$ |
| Esters (ethyl butyrate) | $0.3 \, \mu l/m^3$ |
| Inorganic acids (sulphuric acid) | $10 \, pg$ |
| Pesticides (sulfotep) | $0.1 \, \mu l/m^3$ |
| Terpenes (linalool) | $0.5 \, \mu l/m^3$ |
| Organometallics (trimethylarsine) | $1 \, nl/m^3$ |
| Amides (dimethylformamide) | $20 \, nl/m^3$ |
| Amines (pyridine) | $2 \, nl/m^3$ |

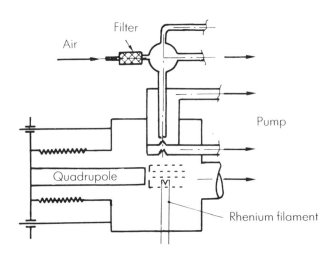

Fig. 18.17 — Apparatus for study of aerosols. (Reproduced by permission, from M. P. Sinha, R. M. Plate, V. L. Vilker and S. K. Friendlander, *Intern. J. Mass Spectrom. Ion Processes* , 1984, **57**, 125. Copyright 1984, Elsevier, Amsterdam.)

nation of ammonia resulting from the metabolism of proteins [9]. The detection limit is of the order of parts in $10^{12}$ v/v.

Another application has been the identification of sulphur compounds in solvent-refined coal liquids. A sample of distillate fraction is first treated chemically with calcium metal in a mixed alkylamine medium to reduce the aromatic sulphur compounds, by cleavage of the carbon–sulphur bond to produce thiophenols. After chemical work-up, the mixture of products is introduced into a triple quadrupole mass spectrometer, the second quadrupole of which is used as a collisional reaction cell in which isobutane (as reaction gas) produces negative ion chemical ionization of the sulphur compounds of interest. The compounds thus identified in the original distillate fraction include alkylbenzothiophenes, dibenzothiophenes and alkylbenzothiophene sulphones [10].

## 18.6  THE STATE OF THE ART

Mass spectrometry is now almost 80 years old, and as already shown in these pages has developed at an ever-increasing rate, from the simplicity of its beginnings to the extremely complex and expensive instrumentation of today. A great deal of this expansion has taken place in the few years since the first appearance of this text in France, and mass spectrometry is now not merely a scientific discipline in its own right, but a conglomeration of multi-disciplinary specializations. Mass spectrometry is used in all branches of chemistry, in physics, geology, environmental, agricultural and space research, the clinical, medical, biological and biochemical fields, and so on. It is clearly beyond the scope of this text to do more than indicate some of the more important of these developments, and the interested reader is urged to make use of the lists of further reading and the general bibliography (given on pages 173–178) to discover the range and ingenuity of modern mass spectrometry in all its aspects. The fact that the four *Analytical Chemistry* biennial reviews of mass spectrometry published in the last three years occupied nearly 140 closely printed pages, but still covered only part of the field, speaks for itself. Only a few representative references will be given in the brief survey below, and for a proper appreciation of the current state of development of mass spectrometry, the reader should browse through these four review papers [11–14].

As indicated in the titles and contents of these reviews, it is convenient to consider mass spectrometry to have three main fields of study — atomic ions, molecular ions and cluster ions. The first will be mainly concerned with inorganic chemistry, and the second with organic chemistry, with a certain amount of overlap. In both, however, there have been major developments in ionization and desorption methods, mass analyser design, detection, and coupling with other techniques such as gas, liquid and supercritical fluid chromatography.

In atomic mass spectrometry the main developments have been in methods of ionization and desorption and in the range of application of the methods. Laser-based techniques, such as laser microprobe analysis [15–24] (which has been applied to cluster analysis [25,26]) and resonant laser ionization [27–35], are now widely exploited, and the latter [36,37], like accelerator mass spectrometry [38–46] may offer the possibility of single-atom detection, as does photon-burst mass spectrometry [47]. Inductively coupled plasmas are extensively used as ionization sources [48–59], and their theory has been examined [60–62]. Thermal ionization is used in isotope dilution analysis [63–71] and there has been a resurgence of interest in glow discharge work [72–77]. Secondary ion mass spectrometry [78–84] and ion microscopy [85–87] continue to be developed, and there have been advances in solving the problem of accuracy and precision in spark source mass spectrometry [88].

In molecular mass spectrometry, the most striking advances are probably those in the field of macromolecules and large biomolecules, with new methods of intact ionization, and of detection. Helium is a good reagent gas for collisionally induced decomposition of large biomolecules [89]. Laser matrix methods [90], two-step laser desorption/ionization [91], and various laser-based techniques in combination with

Fourier transform ion cyclotron resonance [92–96] or multiphoton ionization [97–107] have been used in this connection. High energy collision-induced fragmentation [108–110] has proved fruitful, especially in conjunction with multichannel array detectors [111]). Mass spectrometry coupled with gas chromatography [112,113] and liquid chromatography [114,115] is ever increasingly used, and recently the combination with supercritical fluid chromatography has aroused wide interest [103,116–118]. Ion-trap spectrometers [121,122], time-of-flight analysers [123] and tandem methods [124–126] are all employed in various fields. There have been developments in chemical ionization [127–129], field desorption/ionization [130] and FAB [131]. Neutralization/reionization [132,133] has been used for novel gas phase chemistry [134]. Fourier [135,136] and Hadamard [137] transform methods have been applied. Electrospray [138,139] has been combined with collision-induced dissociation [140]. There is renewed interest in negative-ion chemistry [127,141–143]. Liquid secondary ion mass spectrometry has been the subject of much research [144–146]. Chemometrics and computerization continue to be applied, mainly for pattern recognition [147–157] and library searching [158–161], but also for structure elucidation [162].

Work on ion clusters has covered several topics, such as cluster-ion decay and the associated release of kinetic energy [163–166], photodissociation of size-selected clusters [167–172], dissociation energy measurements [164], and the threshold electron photoionization coincidence (TEPICO) method for fragmentation of heterogeneous clusters [173]. Doubly charged transition-metal clusters have been examined [174,175]. Photoelectron spectroscopy has been used to study the electronic energy level structure of clusters [176–181], and electron energy loss spectroscopy has been used for studying the electronic structure of van der Waals clusters [182]. Other work on van der Waals clusters has included the size dependence of their transformation into metallic clusters [180,183,184], and studies on electron attachment to van der Waals clusters [185–189]. Various aspects of cluster ion research have been reviewed [190–193].

This brief selection of topics should be ample to illustrate the extraordinary breadth and depth of research in mass spectrometry, and the literature on applications is even more voluminous. Perhaps the most fitting comment is that made by Koppenaal in the conclusion section of his 1990 review of atomic mass spectrometry [12].

**REFERENCES**

[1] C. J. Porter, J. H. Beynon and T. Ast, *Org. Mass Spectrom.*, 1981, **16**, 101.
[2] A. Maquestiau, Y. van Haverbeke, R. Flammang, H. Mispreuve, M. Kaisin, J. C. Braekman, D. Daloze and B. Tursch, *Steroids*, 1978, **31**, 31.
[3] M. Bortolotto, J. C. Braekman, A. Daloze and B. Tursch, *Bull. Soc. Chim. Belg.*, 1978, **87**, 539.
[4] G. Vidal, L. Lacoste, J. Alais, A. Lablache-Combier, A. Maquestiau, Y. van Haverbeke, R. Flammang and H. Mispreuve, *Phytochem.*, 1979, **18**, 1405.
[5] J. Fayret, L. Lacoste, J. Alais, A. Lablache-Combier, A. Maquestiau, Y. van Haverbeke, R. Flammang and H. Mispreuve, *Phytochem.*, 1979, **18**, 431.
[6] J. Alais, A. Lablache-Combier, S. Andrieu, A. Maquestiau, Y. van Haverbeke, R. Flammang and H. Mispreuve, *Bull. Mycol. Medicale*, 1979, **8**, 11.
[7] T. L. Kruger, R. G. Cooks, J. L. McLaughlin and R. L. Ranieri, *J. Org. Chem.*, 1977, **42**, 4161.
[8] R. J. Molyneux, A. E. Johnson, J. N. Roitman and M. Benson, *J. Agri. Food. Chem.*, 1979, **27**, 494.
[9] A. M. Lovett, N. M. Reid, J. A. Buckley, J. B. French and D. M. Cameron, *Biomed. Mass Spectrom.*, 1979, **6**, 91.
[10] K. V. Wood, R. G. Cooks, J. A. Laugal and R. A. Benkeser, *Anal. Chem.*, 1985, **57**, 692.
[11] D. W. Koppenaal, *Anal. Chem.*, 1988, **60**, 113R.

[12]  D. W. Koppenaal, *Anal. Chem.*, 1990, **62**, 303R.
[13]  A. L. Burlingame, D. Maltby, D. H. Russell and P. T. Holland, *Anal. Chem.*, 1988, **60**, 294R.
[14]  A. L. Burlingame, D. S. Millington, D. L. Norwood and D. H. Russell, *Anal. Chem.*, 1990, **62**, 268R.
[15]  F. Adams, J. Verlinden and R. Gijbels, *Proc. 6th Intern. Symp. High-Purity Materials*, 1985, **2**, 1.
[16]  L. Moenke-Blankenburgh, *Prog. Anal. Atom. Spectrosc.*, 1986, **9**, 335.
[17]  A. H. Verbueken, F. J. Bruynseels, R. van Grieken and F. Adams, in *Inorganic Mass Spectrometry*, F. Adams, R. Gijbels and R. van Grieken (eds.), Wiley, New York, 1987, Chapter 5.
[18]  A. H. Verbueken, F. J. Bruynseels and R. van Grieken, *Biomed. Mass Spectrom.*, 1985, **12**, 438.
[19]  A. H. Verbueken, I. van de Vyer, M. de Broe and R. van Grieken, *CRC Crit. Rev. Clin. Lab. Sci.*, 1987, **24**, 263.
[20]  R. Kaufmann, *Microbeam Anal.*, 1986, **21**, 177.
[21]  F. Adams and T. Mauney, *Adv. Mass Spectrom.*, 1986, **10**, 507.
[22]  D. S. Simons, *Appl. Surf. Sci.*, 1988, **31**, 103.
[23]  L. van Vaeck, J. Bennett, W. Lauwers, A. Vertes and R. Gijbels, *Microbeam Anal.*, 1988, **23**, 351.
[24]  L. Moenke–Blankenburg, *Laser Microanalysis*, Wiley, New York, 1989.
[25]  R. W. Linton, I. M. Musselman, F. Bruynseels and D. S. Simons, *Microbeam Anal.*, 1987, **22**, 365.
[26]  F. Bruyseels, P. Otten and R. van Grieken, *J. Anal. At. Spectrom.*, 1988, **3**, 237.
[27]  G. S. Hurst and C. G. Morgan (eds.), *Resonance Ionization Spectroscopy 1986*, Inst. Physics, Bristol, 1987.
[28]  J. D. Fassett, L. J. Moore, J. C. Travis and J. R. DeVoe, *Science*, 1985, **230**, 262.
[29]  N. S. Nogar, S. W. Downey and C. M. Miller, *Spectroscopy*, 1985, **1**, 56.
[30]  T. B. Lucatorto and J. E. Parks (eds.), *Resonance Ionization Spectroscopy 1988*, Inst Physics, Bristol, 1989.
[31]  G. S. Hurst and M. G. Payne, *Principles and Applications of Resonance Ionization Spectroscopy*, Hilger/Inst. Physics, Bristol, 1988.
[32]  D. H. Smith, J. P. Young and R. W. Shaw, *Mass Spectrom. Rev.*, 1989, **8**, 345.
[33]  J. P. Young, R. W. Shaw and D. H. Smith, *Anal. Chem.*, 1989, **61**, 1271A.
[34]  J. P. Young, R. W. Shaw, D. E. Goeringer, D. H. Smith and W. H. Christie, *Anal. Instrum.*, 1988, **17**, 41.
[35]  J. D. Fassett and J. C. Travis, *Spectrochim. Acta*, 1988, **43B**, 1409.
[36]  S. D. Kramer, G. S. Hurst, C. H. Chen, M. G. Payne, S. L. Allman, R. C. Philips, B. E. Lehmann, H. Oeschger and H. H. Loosli, *Nucl. Instrum. Methods Phys. Res.*, 1986, **B17**, 395.
[37]  G. S. Hurst, *Phil. Trans. Roy. Soc.*, 1987, **323A**, 155.
[38]  H. E. Gove, in *Treatise on Heavy Ion Science*, D. A. Bromley (ed.), Vol. 7, Plenum Press, New York, 1985, pp. 431–463.
[39]  D. Elmore and F. M. Phillips, *Science*, 1987, **236**, 543.
[40]  J. C. Rucklidge, A. E. Litherland and L. R. Killus, *J. Trace Microprobe Tech.*, 1987, **5**, 23.
[41]  D. Elmore, *Biol. Trace Elem. Res.*, 1987, **12**, 231.
[42]  H. E. Gove, A. E. Litherland and D. Elmore (eds.), *Proc. 4th Intern. Symp. Accelerator Mass Spectrom., Nucl. Instrum. Methods Phys. Res.*, 1987, **B29**, 1.
[43]  A. E. Litherland, *Phil. Trans. Roy. Soc.*, 1987, **323A**, 5.
[44]  W. Woelfli, *Nucl. Instrum. Methods Phys. Res.*, 1987, **B29**, 1.
[45]  K. W. Allen, *Nucl. Instrum. Methods Phys. Res.*, 1988, **B35**, 273.
[46]  J. F. Sellschop, *Nucl. Instrum. Methods Phys. Res.*, 1987, **B29**, 439.
[47]  W. M. Fairbank, Jr., *Nucl. Instrum. Methods Phys. Res.*, 1987, **B29**, 407.
[48]  D. J. Douglas and R. S. Houk, *Prog. Anal. Atom. Spectrosc.*, 1985, **8**, 1.
[49]  A. L. Gray, *Spectrochim. Acta*, 1985, **40B**, 1525.
[50]  A. L. Gray, *J. Anal. At. Spectrom.*, 1986, **1**, 403.
[51]  R. S. Houk, *Anal. Chem.*, 1986, **58**, 97A.
[52]  G. Horlick, S. H. Tan, M. A. Vaughan and Y. Shao, in *Inductively Coupled Plasmas in Analytical Atomic Spectrometry*, A. Montaser and D. W. Golightly (eds.), VCH, New York, 1987, Chapter 10.
[53]  A. L. Gray, in *Inorganic Mass Spectrometry*, F. Adams, R. Gijbels and R. van Grieken (eds.), Wiley, New York, 1988, Chapter 6..
[54]  A. L. Gray, *Z. Anal. Chem.*, 1986, **324**, 561.
[55]  D. W. Koppenaal, *ICP Inf. Newsl.*, 1988, **14**, 267.
[56]  R. S. Houk and J. J. Thompson, *Mass Spectrom. Rev.*, 1988, **7**, 425.
[57]  G. M. Hieftje and G. H. Vickers, *Anal. Chim. Acta*, 1989, **216**, 1.
[58]  J. Marshall, *Anal. Proc.*, 1988, **25**, 238.
[59]  D. J. Douglas, *Can. J. Spectrosc.*, 1989, **34**, No. 2, 38.
[60]  D. J. Douglas and J. B. French, *J. Anal. At. Spectrom.*, 1988, **3**, 743.
[61]  L. B. Lim, R. S. Houk, M. C. Edelson and K. P. Carney, *J. Anal. At. Spectrom.*, 1989, **4**, 365.
[62]  A. L. Gray, *J. Anal. At. Spectrom.*, 1989, **4**, 371.

[63] K. G. Heumann, *Biomed. Mass. Spectrom.*, 1985, **12**, 477.
[64] K. G. Heumann, *Z. Anal. Chem.*, 1986, **324**, 601.
[65] K. G. Heumann, *Z. Anal. Chem.*, 1986, **325**, 661.
[66] K. G. Heumann, *Comments Inorg. Chem.*, 1987, **6**, 145.
[67] A. P. de Leenheer, M. F. LeFevere, W. E. Lambert and E. S. Colinet, *Adv. Clin. Chem.*, 1985, **24**, 111.
[68] K. G. Heumann, in *Inorganic Mass Spectrometry*, F. Adams, R. Gijbels and R. van Grieken (eds.), Wiley, New York, 1988, pp. 301–376.
[69] J. D. Fassett and P. J. Paulsen, *Anal. Chem.*, 1989, **61**, 643A.
[70] J. D. Fassett, *J. Res. Natl. Bur. Std.*, 1988, **93**, 417.
[71] C. B. Smith. *Nucl. Instrum. Methods Phys. Res.*, 1988, **B35**, 364.
[72] W. W. Harrison and B. L. Bentz, *Prog. Anal. Spectrosc.*, 1988, **11**, 53.
[73] W. W. Harrison, *J. Anal. At. Spectrom.*, 1988, **3**, 867.
[74] N. E. Sanderson, E. Hall, J. Clark, P. Charalambous and D. Hall, *Mikrochim. Acta*, 1987 **I**, 275.
[75] R. J. Guidoboni and F. J. Leipziger, *J. Cryst. Growth*, 1988, **89**, 16.
[76] N. Jakubowski, D. Steuwer and W. Vieth, *Z. Anal. Chem.*, 1988, **331**, 145.
[77] J. W. Coburn, *Thin Solid Film*, 1988, **17**, 65.
[78] R. E. Honig, *Intern. J. Mass. Spectrom. Ion Processes*, 1985, **66**, 31.
[79] A. Benninghoven, F. G. Rudenauer and H. W. Werner, *Secondary Ion Mass Spectrometry: Basic Concepts, Instrumental Aspects, Applications*, Wiley–Interscience, New York, 1987.
[80] A. Benninghoven, R. J. Colton, D. S. Simons and H. W. Werner (eds.), *Secondary Ion Mass Spectrometry*, SIMS V, Springer, Berlin, 1986.
[81] A. Lodding, in *Inorganic Mass Spectrometry*, F. Adams, R. Gijbels and R. van Grieken (eds.), Wiley, New York, 1988, Chapter 4.
[82] A. Benninghoven, A. M. Huber and H. W. Werner (eds.), *Secondary Ion Mass Spectrometry*, *SIMS VI*, Wiley, New York, 1988.
[83] R. G. Wilson, F. A. Stevie and C. W. Magee, *Secondary Ion Mass Spectrometry: A Practical Handbook for Depth Profiling and Bulk Impurity Analysis*, Wiley, New York, 1989.
[84] F. Adams, F. Michiels, M. Moens and P. van Espen, *Anal. Chim. Acta*, 1989, **216**, 25.
[85] M. T. Bernius and G. H. Morrison, *Rev. Sci. Instrum.*, 1987, **58**, 1789.
[86] M. Grasserbauer, *Microchem. J.*, 1988, **38**, 24.
[87] J. C. Vickerman, *Chem. Brit.*, 1987, **23**, 969.
[88] A. Rocholi, K. P. Jochum, H. M. Seufert and S. Medini-Best, *Z. Anal. Chem.*, 1988, **331**, 140.
[89] E. Uggerud and R. J. Derrick, *Z. Naturforsch.*, 1989, **44**, 245.
[90] R. W. Nelson, M. J. Rainbau, D. E. Lohr and D. Williams, *Science*, 1989, **246**, 1585.
[91] R. N. Zare, J. H. Hahn, R. Zenobi, *Bull. Chem. Soc. Japan*, 1988, **61**, 87.
[92] J. T. Brenna, *Microbeam Anal.*, 1989, **24**, 306.
[93] H. Y. So and C. L. Wilkins, *J. Phys. Chem.*, 1989, **93**, 1184.
[94] P. F. Greenwood, H. J. Nakat, G. D. Willett, M. A. Wilson and M. G. Strachan, *Preprint Papers*, *Am. Chem. Soc.*, *Div. Fuel Chem.*, 1989, **34**, 773.
[95] M. P. Chiarelli and M. L. Gross, *Anal. Chem.*, 1989, **61**, 1895.
[96] S. Guan and P. R. Jones, *Rev. Sci. Instrum.*, 1988, **59**, 2573.
[97] M. Karas, U. Bahr, A. Ingendoh and F. Hillenkamp, *Angew. Chem.*, 1989, **101**, 805.
[98] M. Karas and F. Hillenkamp, *Microbeam Anal.*, 1989, **24**, 353.
[99] R. Beavis, J. Lindner, J. Grotemeyer, I. M. Atkinson, F. R. Keene and A. E. W. Knight, *J. Am. Chem. Soc.*, 1988, **110**, 7534.
[100] J. Grotemeyer and E. W. Schlag, *Biomed. Environ. Mass Spectrom.*, 1988, **16**, 143.
[101] K. Walter, J. Lindner, J. Grotemeyer and E. W. Schlag, *Chem. Phys.*, 1988, **125**, 155.
[102] M. Karas, A. Ingendoh, U. Bahr and F. Hillenkanp, *Biomed. Environ. Mass Spectrom.*, 1989, **18**, 841.
[103] D. M. Lubman, *Mass Spectrom. Rev.*, 1988, **7**, 535, 559.
[104] U. Boesl, J. Grotemeyer, K. Walter and E. W. Schlag, *Anal. Instrum.*, 1987, **16**, 151.
[105] J. Grotemeyer, U. Boesl, K. Walter and E. W. Schlag, *J. Am. Chem. Soc.*, 1986, **108**, 4233.
[106] J. Grotemeyer, U. Boesl, K. Walter and E. W. Schlag, *Org. Mass Spectrom.*, 1986, **21**, 645.
[107] J. Grotemeyer and E. W. Schlag, *Acc. Chem. Res.*, 1989, **22**, 399.
[108] K. Sato, T. Asada, M. Ishihara, F. Kunihiro, Y. Kammei, E. Kubota, C. E. Costello, S. A. Martin, H. A. Scobie and K. Biemann, *Anal. Chem.*, 1987, **59**, 1652.
[109] R. K. Boyd, P. A. Bott, B. R. Beer, D. J. Harvan and J. R. Hass, *Anal. Chem.*, 1987, **59**, 189.
[110] S. A. Martin and K. Biemann, *Intern. J. Mass Spectrom. Ion Processes*, 1987, **78**, 213.
[111] J. S. Cottrell and S. Evans, *Anal. Chem.*, 1987, **59**, 1990.
[112] F. W. Karasek and R. E. Clement, *Basic Gas Chromatography–Mass Spectrometry: Principles and Techniques*, Elsevier, Amsterdam, 1988.
[113] R. E. Clement, F. I. Onuska, G. A. Eiceman and H. H. Hill, Jr., *Anal. Chem.*, 1988, **60**, 279R; 1990, **62**, 414R.

[114] R. E. Synovec, E. L. Johnson, L. K. Moore and C. N. Benn, *Anal. Chem.*, 1990, **62**, 357R.
[115] A. L. Yergey, C. G. Edmonds, I. A. S. Lewis and M. L. Vestal, *Liquid Chromatography/Mass Spectrometry*, Plenum Press, New York, 1990.
[116] A. J. Berry, D. E. Games, I. C. Mylchreest, J. R. Perkins and S. Pleasance, *JHRC&CC*, 1988, **11**, 61.
[117] D. E. Games, A. J. Berry, I. C. Mylchreest, J. R. Perkins and S. Pleasance, *Anal. Proc.*, 1987, **24**, 371.
[118] T. L. Chester and J. D. Pinkston, *Anal. Chem.*, 1990, **62**, 394R.
[119] R. E. Kaiser, Jr., R. G. Cooks, J. E. P. Syka and G. C. Stafford, Jr., *Rapid Commun. Mass Spectrom.*, 1990, **4**, 30.
[120] J. Allison, *Anal. Chem.*, 1987, **59**, 1072A.
[121] J. M. Louris, J. S. Brodbelt and R. G. Cooks, *Intern. J. Mass Spectrom. Ion Processes*, 1987, **75**, 345.
[122] B. D. Nourse and R. G. Cooks, *Anal. Chim. Acta*, 1990, **228**, 1.
[123] R. J. Cotter, *Biomed. Environ. Mass Spectrom.*, 1989, **18**, 513.
[124] C. K. Meng, M. Mann and J. B. Fenn, *Z. Phys. D.*, 1988, **10**, 361.
[125] P. A. Snyder, C. W. Cross, S. A. Liebman, G. A. Eiceman and R. A. Yost, *Anal. Appl. Pyrolysis*, 1989, **16**, 191.
[126] M. A. Quillan, B. A. Thomson, G. J. Scott and K. Siu, *Rapid Commun. Mass Spectrom.*, 1989, **3**, 145.
[127] P. Kebarle and S. Chowdhury, *Chem. Rev.*, 1987, **87**, 513.
[128] H. Budzikiewicz, *Mass Spectrom. Rev.*, 1986, **5**, 345.
[129] J. B. Westmore and M. M. Alauddin, *Mass Spectrom. Rev.*, 1986, **5**, 381.
[130] R. P. Latimer and H. R. Schulten, *Anal. Chem.*, 1989, **61**, 1201A.
[131] J. A. Sunner, R. Kulatunga and P. Kebarle, *Anal. Chem.*, 1986, **58**, 1312, 2009; 1987, **59**, 1378.
[132] R. Feng, C. Wesdemiotis and F. W. McLafferty, *J. Am. Chem. Soc.*, 1987, **109**, 6521.
[133] J. K. Terlouw and H. Schwarz, *Angew. Chem. Engl. Ed.*, 1987, **26**, 805.
[134] C. Wesdemiotis and F. W. McLafferty, *Chem. Rev.*, 1987, **87**, 485.
[135] F. W. McLafferty and I. Amster, *Intern. J. Mass Spectrom. Ion Processes*, 1986, **72**, 85.
[136] D. F. Hunt, J. Shabanowitz and J. R. Yates, III, *J. Chem. Soc. Chem. Commun.*, 1987, 548.
[137] F. W. McLafferty, D. B. Stauffer, S. Y. Loh and E. R. Williams, *Anal. Chem.*, 1987, **59**, 2212.
[138] R. D. Smith, J. A. Loo, C. J. Barinaga, C. G. Edmonds and H. R. Udseth, *J. Chromatog.*, 1989, **480**, 211.
[139] R. D. Smith and C. J. Barinaga, *Rapid Commun. Mass Spectrom.*, 1990, **4**, 54.
[140] R. D. Smith, J. A. Loo, C. G. Edmonds, C. J. Barinaga and H. R. Udseth, *Anal. Chem.*, 1990, **62**, 882.
[141] D. M. Wetzel and J. I. Brauman, *Chem. Rev.*, 1987, **87**, 607.
[142] R. R. Squires, *Chem. Rev.*, 1987, **87**, 623.
[143] J. H. Bowie, *Mass Spectrom. Rev.*, 1987, **9**, 172.
[144] S. J. Pachuta and R. G. Cooks, *Chem. Rev.*, 1987, **87**, 647.
[145] W. V. Logan, Jr. and S. B. Dorn, *Intern. J. Mass Spectrom. Ion Processes*, 1987, **78**, 99.
[146] E. De Pauw, *Mass Spectrom. Rev.*, 1986, **5**, 191.
[147] W. Windig, W. H. McClennen, H. Stolk and H. L. C. Muezelaar, *Opt. Eng.*, 1986, **25**, 117.
[148] W. J. Dunn, S. L. Emery, W. G. Glen and D. R. Scott, *Environ. Sci. Technol.* 1989, **23**, 1499.
[149] D. R. Scott, *Chemom. Intell. Lab. Syst.*, 1988, **4**, 47.
[150] D. R. Scott, W. J. Dunn and S. L. Emery, *J. Res. Natl. Bur. Std.*, 1988, **93**, 281.
[151] S. A. Howell, W. C. Noble and A. I. Mallet, *Rapid Commun. Mass Spectrom.*, 1989, **3**, 230.
[152] J. van der Greef, A. C. Tas and M. C. ten Noever de Brauw, *Biomed. Environ. Mass Spectrom.*, 1987, **16**, 45.
[153] N. B. Vogt and C. E. Sjoegren, *Anal. Chim. Acta*, 1989, **222**, 135.
[154] N. Simmelt and H. R. Schulten, *J. Anal. Appl. Pyrolysis*, 1988, **15**, 3.
[155] N. Simmelt and H. R. Schulten, *Anal. Chim. Acta*, 1989, **223**, 371.
[156] R. Valarce and G. G. Smith, *J. Anal. Appl. Pyrolysis*, 1988, **15**, 357.
[157] B. Lindner and U. Seydel, *Microbeam Anal.*, 1989, **24**, 286.
[158] D. A. Lauda, Jr., J. R. Cooper and C. L. Wilkens, *Anal. Chem.*, 1986, **58**, 1213.
[159] C. P. Wang and T. L. Isenhour, *Anal. Chem.*, 1987, **59**, 649.
[160] J. T. Clerc, E. Pretsch and M. Zürcher, *Mikrochim. Acta*, 1988 **II**, 217.
[161] J. R. Cooper and C. L. Wilkins, *Anal. Chem.*, 1989, **61**, 1571.
[162] C. G. Enke, A. P. Wade, P. T. Palmer and K. J. Hart, *Anal. Chem.*, 1987, **59**, 1363A.
[163] T. D. Mark, *Intern. J. Mass Spectrom. Ion Processes*, 1987, **79**, 1.
[164] C. Bréchignac, Ph. Cahuzac, J. Leygnier and J. Weiner, *J. Chem. Phys.*, 1989, **90**, 1492.
[165] C. Lifshitz and F. Louage, *J. Phys. Chem.*, 1989, **93**, 5633.

[166] C. A. Woodward, J. E. Upham and A. J. Stace, *Chem. Phys. Lett.*, 1989, **158**, 417.
[167] U. Buck, *J. Phys. Chem.*, 1988, **92**, 1023.
[168] U. Buck, N. J. Gu, Ch. Lauenstein and A. Rudolph, *J. Phys. Chem.*, 1988, **92**, 5561.
[169] U. Buck and Ch. Lauenstein, *J. Chem. Phys.*, 1990, **92**, 4250.
[170] R. Alrichs, S. Brode, U. Buck, M. DeKieviet, Ch. Lauenstein, A. Rudolph and B. Schmidt, *Z. Phys. D*, 1990, **15**, 341.
[171] U. Buck, N. J. Gu, M. Hobein and Ch. Lauenstein, *Chem. Phys. Lett.*, 1989, **163**, 455.
[172] U. Buck, J. Kesper, Ch. Lauenstein, M. Tolle and N. Winter, *Z. Phys. D*, 1989, **12**, 293.
[173] B. Holub-Krappe, G. Ganteför, G. Bröker and A. Ding, *Z. Phys. D*, 1986, **10**, 314.
[174] T. Leisner, O. Echt, C. Kandler, D. Kreisle and E. Recknagel, *Intern. J. Mass Spectrom. Ion Processes*, 1989, **87**, R19.
[175] C. Bréchignac, Ph. Cahuzac, F. Carlier and J. Leygnier, *Phys. Rev. Lett.*, 1989, **63**, 1368.
[176] G. Ganteför, M. Gausa, K. H. Meiwes-Broer and H. O. Lutz, *Z. Phys. D*, 1988, **9**, 253.
[177] G. Ganteför, M. Gausa, K. H. Meiwes-Broer and H. O. Lutz, *Faraday Discuss. Chem. Soc.*, 1988, **86**, 197.
[178] D. G. Leopold, J. H. Ho and W. C. Lineberger, *J. Chem. Phys.*, 1987, **86**, 1715.
[179] C. L. Pettiette, S. H. Yang, M. J. Craycraft, J. Conceicao, R. T. Laaksonen, O. Cheshnovsky and R. E. Smalley, *J. Chem. Phys.*, 1988, **88**, 5377.
[180] K. Rademann, B. Kaiser, U. Even and F. Heusel, *Phys. Rev. Lett.*, 1987, **59**, 2319.
[181] G. Ganteför, M. Gausa, K. H. Meiwes-Broer and H. O. Lutz, *Z. Phys. D.*, 1989, **12**, 405.
[182] A. Burose, C. Becker and A. Ding, in *Proc. Symp. Atomic and Surface Physics, Innsbruck 18–24 March* 1990, T. D. Märk (ed.), in the press.
[183] C. Bréchignac, M. Broyer, Ph. Cahuzac, G. Delacretaz, P. Labastie, J. P. Wolf and L. Wöste, *Phys. Rev. Lett.*, 1988, **60**, 275.
[184] H. Haberland, H. Kornmeier, H. Langosch, M. Oschwald and G. Tanner, *J. Chem. Soc. Faraday Trans.*, II, 1990, **86**, 2473.
[185] A. Stamatovic, *Electron attachment to van der Waals clusters*, in *Electronic and Atomic Collisions*, H. G. Gilbody, W. R. Newell, F. H. Read and A. C. H. Smith (eds.), Elsevier, Amsterdam, 1988.
[186] T. Kondow, T. Nagata and K. Kuchitu, *Z. Phys. D*, 1989, **12**, 291.
[187] A. Stamatovic, P. Scheier and T. D. Märk, *J. Chem. Phys.*, 1988, **88**, 6884.
[188] H. Haberland, C. Ludewight and T. Richter, *Z. Phys. D*, 1989, **12**, 289.
[189] T. Kraft, M. W. Ruf and H. Hotop, *Z. Phys. D.*, 1989, **14**, 179.
[190] J. Jortner, U. Even, N. Ben-Horin, D. Scharf, R. N. Barnett and U. Landman, *Z. Phys. D*, 1989, **12**, 167.
[191] T. Kondow, *J. Phys. Chem.*, 1987, **91**, 1307.
[192] T. D. Märk, *Adv. Mass Spectrom.*, 1985, **10**, 379.
[193] T. D. Märk and P. Howorka, in *Proc. Symp. Atomic and Surface Physics, Innsbruck 18–24 March* 1990, T. D. Märk (ed.), in the press.

# Further reading

**Chapter 9**

W. C. M. C. Kokke, S. Epstein, S. A. Look, G. H. Rau, W. H. Fenical and C. Djerassi, *J. Biol. Chem.*, 1984, **259**, 8168.

J. R. Turnlund and P. E. Johnson (eds.), *Stable Isotopes in Nutrition*, Am. Chem. Soc., Washington, DC, 1984.

**Chapter 10**

P. J. Derrick, A. M. Falick and A. L. Burlingame, *J. Am. Chem. Soc.*, 1972, **94**, 6794.

J. L. Holmes, *Org. Mass Spectrom.*, 1985, **20**, 169.

C. K. Mak and R. S. Tse, *Org. Mass Spectrom.*, 1984, **19**, 227.

S. C. Unger, A. Vincze, R. G. Cooks, R. Chrisman and L. D. Rothman, *Anal. Chem.*, 1981, **53**, 976.

**Chapter 11**

E. W. McDaniel, V. Čermák, A. Dalgarno, E. E. Ferguson and L. Friedman, *Ion–Molecule Reactions*, Wiley–Interscience, New York, 1970.

E. Constantin, *Org. Mass Spectrom.*, 1982, **17**, 346.

I. Howe, in *Mass Spectrometry (Specialist Periodical Report)*, Vol. 7, Royal Society of Chemistry, London, 1984.

M. T. Bowers (ed.), *Gas Phase Ion Chemistry*, Academic Press, New York, 1972.

D. E. Magnoli and J. R. Murdoch, *J. Am. Chem. Soc.*, 1981, **103**, 7465.

C. E. Hudson and D. J. McAdoo, *Org. Mass Spectrom.*, 1982, **17**, 366.

C. Westdemiotis and F. W. McLafferty, *Org. Mass Spectrom.*, 1981, **16**, 381.

W. R. Davidson, M. T. Bowers, T. Su and D. H. Aue, *Intern. J. Mass Spectrom. Ion Phys.*, 1977, **24**, 83.

L. N. Morgenthaler and J. Eyler, *Intern. J. Mass Spectrom. Ion. Phys.*, 1981, **37**, 153.

S. A. Safron, in *Mass Spectrometry (Specialist Periodical Reports)*, Vol. 7, Royal Society of Chemistry, London, 1984.

R. V. Hodges, P. B. Armentrout and J. L. Beauchamp, *Intern. J. Mass Spectrom. Ion Phys.*, 1979, **29**, 375.

J. L. Beauchamp, A. E. Stevens and R. R. Corderman, *Pure Appl. Chem.*, 1979, **51**, 467.

S. Matsuoka, H. Nakamura and T. Tamura, *Intern. J. Mass Spectrom. Ion Phys.*, 1981, **37**, 315.

R. Thomas, J. Barassin and A. Barassin, *Intern. J. Mass Spectrom. Ion Phys.*, 1981, **41**, 95.

J. L. McCrumb and P. Warneck, *Intern. J. Mass Spectrom. Ion Phys.*, 1981, **37**, 127.

**Chapter 12**

R. Rechsteiner, R. Monot, L. Wöste, J. M. Zellweger and H. van den Bergh, *Ber. Früjahr. Schw. Phys. Gesells.*, 1981, **54**, 283; *Helv. Phys. Acta*, 1981, **54**, 282.

W. R. Davidson, M. T. Bowers, T. Su and D. H. Aue, *Intern. J. Mass Spectrom. Ion Phys.*, 1977, **24**, 83.

H. Villinger, J. H. Futrell, A. Saxer, R. Richter and W. Lindinger, *J. Chem. Phys.*, 1984, **80**, 2543.

P. Kebarle, *Ann. Rev. Phys. Chem.*, 1977, **28**, 445.

R. G. Keesee and A. W. Castleman, Jr., *Chem. Phys. Lett.*, 1980, **74**, 139.

J. N. Tang and A. W. Castleman, Jr., *J. Chem. Phys.*, 1972, **57**, 3638.
F. H. Field, *J. Am. Chem. Soc.*, 1969, **91**, 2827.
A. B. Rakshit and P. Warneck, *Intern. J. Mass Spectrom. Ion Phys.*, 1981, **40**, 135.
R. Johnson, C. M. Huang and M. A. Biondi, *J. Chem. Phys.*, 1975, **63**, 3374.
R. J. Beuhler, S. Ehrenson and L. Friedman, *J. Chem. Phys.*, 1983, **79**, 5982.
D. Price and J. F. J. Todd, *Dynamic Mass Spectrometry*, Heyden, London, 1978.
U. Nagashima, H. Shinohara, N. Nishi and H. Tanaka, *J. Chem. Phys.*, 1986, **84**, 209.
H. Böhringer and F. Arnold, *Intern. J. Mass Spectrom. Ion Phys.*, 1983, **49**, 61.


**Chapter 13**
R. D. Smith, J. A. Loo, C. G. Edmonds, C. J. Barinaga and H. R. Udseth, *Anal. Chem.*, 1990, **62**, 882.
H. R. Morris (ed.), *Soft Ionizaton Biological Mass Spectrometry*, Royal Society of Chemistry, London, 1980.
B. T. Chait, W. C. Agosta and F. H. Field, *Intern. J. Mass Spectrom. Ion Phys.*, 1981, **39**, 339.
G. J. Q. van der Peyl, J. Haverkamp and P. G. Kistemaker, *Intern. J. Mass Spectrom. Ion Phys.*, 1982, **42**, 125.
G. J. Q. van der Peyl, K. Isa, J. Haverkamp and P. G. Kistemaker, *Nucl. Instrum. Meth.*, 1982, **198**, 125; *Org. Mass Spectrom.*, 1981, **16**, 416.
H.-Y. Kim and N. Salem, *Anal. Chem.*, 1986, **58**, 9.
D. J. Liberato and A. L. Yergey, *Anal. Chem.*, 1986, **58**, 6.
D. E. Games, *Biomed. Mass Spectrom.*, 1981, **8**, 454.
M. Tsuchiya and H. Kawabara, *Anal. Chem.*, 1984, **56**, 14.
M. Barber, R. S. Bordoli, R. Sedgwick and A. N. Tyler, *Nature*, 1981, **293**, 279.
T. T. Chang, J. O. Lay, Jr., and R. J. Francel, 1984, **56**, 109.
K. D. Cook and H. W. S. Chan, *Intern. J. Mass Spectrom. Ion Phys.*, 1983, **54**, 135.
C. Bojesen, *Org. Mass Spectrom.*, 1983, **20**, 413.
M. Barber, R. S. Bordoli, G. J. Elliott, R. D. Sedgwick and A. N. Tyler, *Anal. Chem.*, 1982, **54**, 654A.
J. M. Miller and R. Theberge, *Org. Mass Spectrom.*, 1985, **20**, 600.
P. J. Arpino, P. Krien, S. Vajta and G. Devant, *J. Chromatog.*, 1981, **203**, 117.
P. J. Arpino, J. P. Bounine, M. Dedieu and G. Guiochon, *J. Chromatog.*, 1983, **271**, 43.
J. de Graeve, F. Berthou and M. Prost, *Méthodes Chromatographiques Couplées à un Spectromètre de Masse*, Masson, Paris, 1986.


**Chapter 14**
E. W. McDaniel, V. Čermák, A. Dalgarno, E. E. Ferguson and L. Friedman, *Ion–Molecule Reactions*, Wiley, New York, 1970.
D. W. Fahey, F. C. Fehsenfeld and E. E. Ferguson, *Geophys. Res. Lett.*, 1981, **8**, 1115.
R. A. Perry, B. R. Rowe, A. A. Viggiano, D. L. Albritton, E. E. Ferguson and F. C. Fehsenfeld, *Geophys. Res. Lett.*, 1980, **7**, 693.
M. L. Huertas, A. M. Marty and J. Fontan, *J. Geophys. Res.*, 1974, **79**, 1737.
M. L. Heurtas, *Atm. Environ.*, 1975, **9**, 1018.
S. Matsuoka and T. Tamura, *Intern. J. Mass Spectrom. In. Phys.*, 1981, **40**, 331.
W. B. Hanson, S. Sanatani and J. H. Hoffman, *J. Geophys. Res.*, 1981, **86**, 7210.
R. W. Kiser, *Introduction to Mass Spectrometry and its Applications*, Prentice-Hall, Englewood Cliffs, NJ, 1965.
F. Arnold, J. Kissel, D. Krankowsky, H. Wieder and J. Zähringer, *J. Atm. Ter. Phys.*, 1971, **33**, 1169.
R. S. Narcisi, A. D. Bailey, L. D. Lucca, C. Sherman and D. M. Thomas, *J. Atm. Ter. Phys.*, 1971, **33**, 1147.
F. Arnold, R. Fabian, E. E. Ferguson and W. Joos, *Planet. Space Sci.*, 1981, **29**, 195.
F. Arnold, G. Henschen and E. E. Ferguson, *Planet Space Sci.*, 1981, **29**, 185.
F. Arnold and G. Henschen, *Nature*, 1978, **275**, 521.
D. K. Bohme, *NATO ASI Ser. Ser. C.*, 1984, No. 118, 111.


**Chapter 15**
R. G. Cooks, J. H. Beynon, R. M. Caprioli and G. R. Lester, *Metastable Ions*, Elsevier, Amsterdam, 1973.
H. M. Rosenstock, R. Stockbauer, and A. C. Parr, *Intern. J. Mass Spectrom. Ion Phys.*, 1978, **27**, 185.
R. Stockbauer and H. M. Rosenstock, *Intern. J. Mass Spectrom. Ion Phys.*, 1981, **38**, 323.
C. E. Klots, *J. Phys. Chem.*, 1971, **75**, 1526.
C. E. Klots, *J. Chem. Phys.*, 1973, **58**, 5364.
P. Ausloos, *J. Am. Chem. Soc.*, 1981, **103**, 3931.

C. Lifshitz, *Mass Spectrom. Rev.*, 1982, **1**, 309.

J. C. Lorquet, *Org. Mass Spectrom.*, 1981, **16**, 469.

R. Bombach, J. Dannacher, J. P. Stadelman, J. Vogt, L. R. Thorne and J. L. Beauchamp, *Chem. Phys.*, 1982, **66**, 403.

C. J. Cerjan and H. W. Miller, *J. Chem. Phys.*, 1981, **75**, 2800.

J. J. Butler and T. Baer, *J. Am. Chem. Soc.*, 1980, **102**, 6764.

J. J. Butler, M. L. Fraser-Monteiro, L. Fraser-Monteiro, T. Baer and J. R. Haas, *J. Phys. Chem.*, 1982, **86**, 747.

C. J. Proctor, B. Kralj, A. G. Brenton and J. H. Beynon, *Org. Mass Spectrom.*, 1980, **15**, 619.

J. C. Lorquet, in *Mass Spectrometry (Specialist Periodical Report)*, Vol. 7, Royal Society of Chemistry, London, 1984.

I. Powis, in *Mass Spectrometry (Specialist Periodical Report)* Vol. 7, Royal Society of Chemistry, 1984.

J. H. Beynon and J. G. Gilbert, *Application of Transition State Theory to Unimolecular Reactions*, Wiley, Chichester, 1984.

**Chapter 18**

G. M. Hieftje, J. C. Travis and F. E. Lytle (eds.), *Lasers in Chemical Analysis*, Humana Press, Clifton, NJ, 1981.

F. W. McLafferty (ed.), *Tandem Mass Spectrometry*, Wiley, New York, 1983.

R. G. Cooks, J. H. Beynon, R. M. Caprioli and G. R. Lester, *Metastable Ions*, Elsevier, Amsterdam, 1973.

C. J. McNeal and R. D. Macfarlane, *J. Am. Chem. Soc.*, 1981, **103**, 1609.

M. Barber, R. S. Bordoli, G. J. Elliott, R. D. Sedgwick and A. N. Tyler, *Anal. Chem.*, 1982, **54**, 654 A.

M. Barber, R. S. Bordoli, R. D. Sedgwick and A. N. Tyler, *Nature*, 1981, **293**, 270.

F. Österberg and K. Lindström, *Org. Mass Spectrom.*, 1985, **20**, 515.

C. K. Mak and R. S. Tse, *Org. Mass Spectrom.*, 1984, **19**, 227.

M. P. Sinha, R. M. Platz, V. L. Vilker and S. K. Friedlander, *Intern. J. Mass Spectrom. Ion Processes* , 1984, **57**, 125.

A. M. Lovett, N. M. Reid, J. A. Buckley, J. B. French and D. M. Cameron, *Biomed. Mass Spectrom.*, 1979, **6**, 91.

K. D. Cook and K. W. S. Chan, *Intern. J. Mass Spectrom. Ion Phys.*, 1983, **54**, 135.

C. L. Wilkins, D. A. Weil, C. L. C. Yang and C. F. Ijames, *Anal. Chem.*, 1985, **57**, 520.

T. B. Simpson, J. G. Black, I. Burak, E. Yablonovich and N. Bloembergen, *J. Chem. Phys.*, 1985, **83**, 628.

T. Matsuo, H. Matsuda, I. Katakuse, Y. Wada, T. Fujita and A. Hayashi, *Biomed. Mass Spectrom.*, 1981, **8**, 25.

J. L. Holmes, *Org. Mass Spectrom.*, 1985, **20**, 169.

R. D. Macfarlane and D. F. Torgerson, *Science*, 1976, **191**, 920.

C. J. McNeal, *Anal. Chem.*, 1982, **54**, 43A.

C. J. McNeal, K. K. Ogilvie, N. Y. Theriault and M. J. Nemer, *J. Am. Chem. Soc.*, 1982, **104**, 981.

M. Suzuki, *Biomed. Mass Spectrom.*, 1983. **10**, 352.

H. R. Schulten, *Intern. J. Mass Spectrom. Ion Phys.*, 1979, **32**, 97.

B. Calas, J. Méry, J. Parello, J. C. Promé, J. Roussel and D. Patouraux, *Biomed. Mass Spectrom.*, 1980, **7**, 288.

H. M. Schiebel and H. R. Schulten, *Z. Naturforsch.*, 1981, **36b**, 967.

E. Constantin and R. Hueber, *Org. Mass Spectrom.*, 1982, **17**, 460.

E. Constantin, Y. Nakatani, G. Teller, R. Hueber and G. Ourisson, *Bull. Soc. Chim. France*, 1981, II-303.

E. Constantin, B. Dietrich, M. W. Dietrich, J. M. Lehn and R. B. Session, in *Recent Developments in Mass Spectrometry in Biochemical and Environmental Research*, A. Frigerio (ed.), Elsevier, Amsterdam, 1983.

# Reviews and articles of interest

H. R. Schulten, *Intern. J. Mass Spectrom. Ion Phys.*, 1979, **32**, 97.

C. Brunnee, *Intern. J. Mass Spectrom. Ion Processes*, 1987, **76**, 121.

*Royal Society of Chemistry Specialist Reports on Mass Spectrometry*, Vol. 1, 1971; Vol. 2, 1973; Vol. 3 1975; Vol. 4, 1977; Vol. 5, 1979; Vol. 6, 1981; Vol. 7, 1984; Vol. 8, 1985; Vol. 9, 1987; Vol. 10, 1989.

*Mass Spectrometry Reviews*, 1982–, Wiley, New York.

*Advances in Mass Spectrometry* (Conference Proceedings), Institute of Petroleum, London and various publishers.

A. M. Ure, A. T. Ellis and J. G. Williams, *J. Anal. At. Spectrom.*, 1988, **3**, 175R.

J. R. Bacon, A. T. Ellis and J. G. Williams, *J. Anal. At. Spectrom.*, 1989, **4**, 199R.

R. S. Houk, *Anal. Chem.*, 1986, **58**, 97A (ICP/MS).

W. W. Harrison, K. R. Hess, R. K. Marcus and F. L. King, *Anal. Chem.*, 1986, **58**, 341A (Glow discharge MS).

*Anal. Chem.*, 1986, **58**, 536 (Instrumentation for *Galileo* mission to Jupiter).

K. Biemann, *Anal. Chem.*, 1986, **58**, 1288A (Protein sequencing by MS).

T. R. Covey, A. D. Lee, A. P. Bruins and J. D. Henson, *Anal. Chem.*, 1986, **58**, 1451A (LC/MS).

D. M. Lubman, *Anal. Chem.*, 1987, **59**, 31A (Optically selective MS).

R. E. Taylor, *Anal. Chem.*, 1987, **59**, 317A (Dating techniques).

G. W. Small, *Anal. Chem.*, 1987, **59**, 535A (Automated spectral interpretation).

C. L. Wilkins, *Anal. Chem.*, 1987, **59**, 571A (GC/MS).

S. Borman, *Anal. Chem.*, 1987, **59**, 588A (Molecular SIMS).

S. Borman, *Anal. Chem.*, 1987, **59**, 701A (Multichannel detection).

S. Borman, *Anal. Chem.*, 1987, **59**, 769A (GC/MS, LC/MS, MAGIC, SFC/MS).

M. Warner, *Anal. Chem.*, 1987, **59**, 855A (LC/MS, SFC/MS, GC/MS).

J. Allison and R. M. Stepnowski, *Anal. Chem.*, 1987, **59**, 1072A (Ion trapping).

C. G. Enke, A. P. Wade, P. T. Palmer and K. J. Hart, *Anal. Chem.*, 1987, **59**, 1363A (Automated MS/MS structure elucidation).

D. F. Gurka, L. D. Betowski, T. A. Hinners, E. M. Heithmar, R. Titus and J. M. Henshaw, *Anal. Chem.*, 1988, **60**, 454A (Environmental applications of coupled MS techniques).

I. L. Davies, M. W. Raynor, J. P. Kithinji, K. D. Bartle, P. T. Williams and G. E. Andrews, *Anal. Chem.*, 1988, **60**, 683A (GC/MS, LC/MS, SFE/GC).

R. J. Cotter, *Anal. Chem.*, 1988, **60**, 781A (Plasma desorption).

R. D. Smith, B. W. Wright and C. R. Yonker, *Anal. Chem.*, 1988, **60**, 1323A (SFC/MS).

M. Warner, *Anal. Chem.*, 1989, **61**, 101A (Shroud of Turin).

J. D. Fassett and P. J. Paulsen, *Anal. Chem.*, 1989, **61**, 643A (Isotope dilution MS).

R. P. Lattimer and H.-R. Schulten, *Anal. Chem.*, 1989, **61**, 1201A (Field ionization and desorption).

J. P. Young, R. W. Shaw and D. H. Smith, *Anal. Chem.*, 1989, **61**, 1271A (Resonance ionization).

R. P. Caprioli, *Anal. Chem.*, 1990, **62**, 477A (Continuous flow FAB).

J. A. Syage, *Anal. Chem.*, 1990, **62**, 505A (Real-time detection).

E. C. Huang, T. Wachs, J. J. Conboy and J. D. Henion, *Anal. Chem.*, 1990, **62**, 713A (Atmospheric pressure ionization).

W. W. Harrison, C. M. Barshick, J. A. Klingler, P. H. Ratliff and Y. Mel, *Anal. Chem.*, 1990, **62**, 943A (Glow discharge techniques).

# General bibliography

F. H. Field and J. L. Franklin, *Electron Impact Phenomena*, Academic Press, New York, 1957.

H. Hinterberger, *Nuclear Masses and their Determination*, Pergamon Press, London, 1957.

J. H. Beynon, *Mass Spectrometry and its Application to Organic Chemistry*, Elsevier, Amsterdam, 1960.

A. Birkenfeld, G. Haase and H. Zahn, *Massenspektrometrische Isotopenanalyse*, VEB Deutscher Verlag der Wissenchaften, Berlin, 1962.

C. A. McDowell (ed.), *Mass Spectrometry*, McGraw-Hill, New York, 1963.

E. W. McDaniel, *Collisional Phenomena in Ionized Gases*, Wiley, New York, 1964.

R. I. Reed (ed.), *Mass Spectrometry*, Academic Press, New York, 1965.

R. W. Kiser, *Introduction to Mass Spectrometry and its Applications*, Prentice-Hall, Englewood Cliffs, NJ, 1965.

A. J. Ahearn, *Mass Spectrometric Analysis of Solids*, Elsevier, Amsterdam, 1966.

R. I. Reed, *Applications of Mass Spectrometry to Organic Chemistry*, Academic Press, New York, 1966.

H. C. Hill, *Introduction to Mass Spectrometry,* Heyden, London, 1966.

H. Budzikiewicz, C. Djerassi and D. H. Williams, *Mass Spectrometry of Organic Compounds*, Holden-Day, San Francisco, 1967.

J. Roboz, *Introduction to Mass Spectrometry: Instrumentation and Techniques*, Interscience, New York, 1968.

D. Price and J. E. Williams, *Time-of-Flight Mass Spectrometry*, Pergamon Press, Oxford, 1969.

E. W. McDaniel, V. Čermák, A. Dalgarno, E. E. Ferguson and L. Friedman, *Ion–Molecule Reactions*, Wiley, New York, 1970.

R. Binks, J. S. Littler and R. I. Cleaver, *Tables for Use in High Resolution Mass Spectrometry*, Heyden, London, 1970.

R. G. Cooks, J. H. Beynon, R. M. Caprioli and G. R. Lester, *Metastable Ions*, Elsevier, Amsterdam, 1973.

M. R. Litzow, T. R. Spalding, *Mass Spectrometry of Inorganic and Organometallic Compounds*, Elsevier, Amsterdam, 1973.

R. H. Dawson (ed.), *Quadrupole Mass Spectrometry*, Elsevier, Amsterdam, 1976.

H. D. Beckey, *Principles of Field Ionization and Field Desorption Mass Spectrometry*, Pergamon Press, Oxford, 1977.

M. T. Bowers, *Gas Phase Ion Chemistry*, Academic Press, New York, 1979.

B. S. Middleditch (ed.), *Practical Mass Spectrometry: A Contemporary Introduction*, Plenum Press, New York, 1979.

J. R. Chapman, *Computers in Mass Spectrometry*, Heyden, London, 1978.

G. R. Waller and O. C. Dermer (eds.), *Biochemical Applications of Mass Spectrometry*, Wiley, New York, 1980.

F. W. McLafferty, *Interpretation of Mass Spectra*, Univ. Sci. Books, Mill Valley, CA, 1980.

G. M. Heiftje, J. C. Travis and F. E. Lytle (eds.), *Lasers in Chemical Analysis*, Humana Press, Clifton, NJ, 1981.

D. Price and J. F. J. Todd (eds.), *Dynamic Mass Spectrometry*, Heyden, London, 1981.

H. R. Morris (ed.), *Soft Ionization Biological Mass Spectrometry*, Heyden, London, 1981.

J. H. Beynon and M. L. McGlashan (eds.), *Current Topics in Mass Spectrometry and Chemical Kinetics*, Heyden, London, 1982.

F. W. McLafferty (ed.), *Tandem Mass Spectrometry*, Wiley, New York, 1983.

J. H. Beynon and J. C. Gilbert, *Application of Transition State Theory to Unimolecular Reactions*, Wiley, Chichester, 1984.

J. de Graeve, F. Berthou and M. Prost, *Méthodes Chromotographiques Couplées à un Spectromètre de Masse*, Masson, Paris, 1986.

S. J. Gaskell (ed.), *Mass Spectrometry in Biomedical Research*, Wiley, New York, 1986.

C. J. McNeal (ed.), *Mass Spectrometry in the Analysis of Large Molecules*, Wiley, New York, 1986.

A. Benninghoven, F. G. Rudenauer and H. W. Werner, *Secondary Ion Mass Spectrometry: Basic Concepts, Instrumental Aspects, Applications and Trends*, Wiley, New York, 1986.

J. H. Futrell, *Gaseous Ion Chemistry and Mass Spectrometry*, Wiley, New York, 1986.

F. A. White and G. M. Wood, *Mass Spectrometry — Applications in Science and Engineering*, Wiley, New York, 1986.

M. V. Buchanan (ed.), *Fourier Transform Mass Spectrometry*, American Chemical Society, Washington, D.C., 1987.

J. Gilbert (ed.), *Applications of Mass Spectrometry for Food Science*, Elsevier, London, 1987.

A. E. Litherland, K. W. Allen and E. T. Hall, *Ultra-High-Sensitivity Mass Spectrometry with Accelerators*, Royal Society, London, 1987.

F. Adams, R. Gijbels and R. van Grieken, *Inorganic Mass Spectrometry*, Wiley, New York, 1988.

K. L. Busch, K. L. Glish and S. A. McLuckey (eds.), *Mass Spectrometry: Technique and Applications of Tandem Mass Spectrometry*, VCH, New York, 1988.

A. R. Date and A. L. Gray, *Applications of Inductively Coupled Plasma Mass Spectrometry*, Blackie, Glasgow, 1988.

A. M. Lawson (ed.), *Mass Spectrometry*, de Gruyter, Berlin, 1988.

F. W. Karasek and R. E. Clement, *Basic Gas Chromatography–Mass Spectrometry: Principles and Techniques*, Elsevier, Amsterdam, 1988.

L. Moenke-Blankenburg, *Laser Microanalysis*, Wiley, New York, 1989.

L. Prókal, *Field Desorption Mass Spectrometry*, Dekker, New York, 1989.

A. L. Yergey, C. G. Edmonds, I. A. S. Lewis and M. L. Vestal, *Liquid Chromatography/Mass Spectrometry*, Plenum Press, New York, 1990.

C. N. McEwan and B. S. Larsen (eds.), *Mass Spectrometry of Biological Materials*, Dekker, New York, 1990.

C. H. Suelter and J. T. Watson (eds.), *Biomedical Applications of Mass Spectrometry*, Wiley, New York, 1990.

D. M. Desiderio, *Mass Spectrometry of Peptides*, CRC Press, Boca Raton, FL, 1990.

A. Burlingame and J. L. McCloskey (eds.), *Biological Mass Spectrometry*, Elsevier, Amsterdam, in the press.

# Index